HITLER'S ARMY

HITLER'S ARMY

Soldiers, Nazis, and War in the Third Reich

Omer Bartov

New York Oxford
OXFORD UNIVERSITY PRESS
1991

Oxford University Press

Oxford   New York   Toronto
Delhi   Bombay   Calcutta   Madras   Karachi
Petaling Jaya   Singapore   Hong Kong   Tokyo
Nairobi   Dar es Salaam   Cape Town
Melbourne   Auckland

and associated companies in
Berlin   Ibadan

Copyright © 1991 by Oxford University Press, Inc.

Published by Oxford University Press, Inc.
200 Madison Avenue, New York, New York 10016

Oxford is a registered trademark of Oxford University Press

Library of Congress Cataloging-in-Publication Data
Bartov, Omer.
Hitler's army : soldiers, Nazis, and war in the Third Reich /
Omer Bartov.
p. cm.   Includes bibliographical references and index.
ISBN 0-19-506879-3
1. Germany—Armed Forces—History—20h century.
2. Germany—Armed Forces—Political activity.   3. National socialism.
4. World War, 1939–1945—Atrocities.   I. Title.
D757.B27   1991   940.54'0943—dc20   90-48960

9 8 7 6 5 4 3 2 1

Printed in the United States of America
on acid-free paper

For Raz

Preface

This book is not a comprehensive history of the German army and its relationship with regime and society in the Third Reich. It is an essay, arguing four distinct but related theses which, taken together, may enhance our insight into the Nazification of Germany's soldiers. This process began long before the war, and some of its roots predate the Nazi regime. Yet it was during the war, and most importantly on the Eastern Front, that the Wehrmacht finally became Hitler's army. Moreover, as the vast majority of German troops fought for most of the war against the Red Army, one can say that for the average soldier the fighting in Russia constituted the crucial component of his war experience. Consequently, though I refer to the impact of the prewar years on the soldiers' perception of reality, and take note of the rather different experience on other fronts, I intentionally concentrate on that vast confrontation between Germany and the Soviet Union, where the Wehrmacht both won its greatest victories and was finally destroyed, and where the progressive ideological penetration of the army reached its peak, motivating the troops to fight with extraordinary resilience, on the one hand, and to commit unprecedented crimes, on the other. I am mainly concerned here with the land forces, or *Heer*. The involvement of the SS in Nazi policies has already been widely discussed, whereas the experience of the Navy and *Luftwaffe*, which had conscripted a much smaller share of Germany's manpower, was in many ways different, though particularly the pilots flying over Russia were just as exposed to the Nazi view of that war as the soldiers were fighting on the ground. As the basis for my arguments I have used some documents first published in my previous book, *The Eastern Front, 1941–45, German Troops and the Barbarisation of Warfare*, adding to them other unpublished documentary material and utilizing the important secondary works that have appeared in the intervening years. It should also be noted that while my earlier monograph provided a close analysis of three combat divisions on the Eastern Front, this work substantially widens the scope of my argument by asserting that the experience of that front was crucial to the German army as a whole, indeed to German society both during and after the war.

While writing this book I have profited a great deal from lengthy periods of research; I also owe more than can be acknowledged to endless discussions with specialists, students, and friends. The recent *Historikerstreit* has forced me to rethink and reformulate most of my ideas; and the momentous political developments in Germany, Eastern Europe, and the Soviet Union have brought into much sharper relief the contemporary relevance of what many of us had come to see as "mere" history. I must also admit that my personal experiences as an Israeli soldier and citizen have had a substantial, if indirect impact on my views as an historian. If in Israeli political debates I tend to cite the German example, when writing about the Wehrmacht I find myself drawing on my own experiences. Nonetheless, though I have tried to understand the mentality of Hitler's soldiers, I have not felt the need to identify with them. What I have written is intended to contribute to our understanding of how ordinary men can be made into both highly professional and determined soldiers, brutalized instruments of a barbarous policy, and devoted believers in a murderous ideology; how they can be taught to live in an inverted world of fictitous images, and why their distorted view of reality is perpetuated long after the objective conditions which had prompted it have disappeared in a surge of horrendous destruction.

I owe thanks to a large number of individuals and institutions for guidance, ideas, and support. At Stanford University I was launched on my first postgraduate attempts to understand history by Gordon Craig, Peter Paret, Gordon Wright, and Chimen

Abramsky. During my years at Oxford and in subsequent visits I received valuable advice from Tony Nicholls, Tim Mason, and Michael Howard. I would also like to thank Pogge von Strandmann, Richard Bessel, and Volker Berghahn. Other colleagues in Britain who kindly invited me to read papers at various universities in that country and provided me with much constructive criticism are Jeremy Noakes, Eve Rosenhaft, Jill Stephenson, and Dick Geary. I owe a special debt to Ian Kershaw for years of friendship and inestimable assistance. In Germany I am grateful to the staff of the Militärgeschichtliches Forschungsamt in Freiburg, and especially to Wilhelm Deist, Bernd Kroener, Hans Umbreit, Jürgen Förster, Wolfram Wette, and Manfred Messerschmidt. Bernd Wegner, his wife Anneli, and their three children have given me a home in Germany to which I will always hope to return. The staff of the Bundesarchiv-Militärarchiv guided me with great expertise through the maze of documents. Thanks is also due to Reinhard Rürup for inviting me to spend a constructive summer as a guest of the Technische Hochschule in Berlin, and to the Friedrich Meinecke Library for its rich collection of secondary works. Cornelia Essner's friendship and hospitality made my stay in Berlin much more enjoyable than it would have otherwise been. Hans Ulrich-Wehler's interest in my work has greatly encouraged me. At Tel-Aviv University I wish to thank all my colleagues at the Department of History. I am particularly indebted to Zvi Yavetz, Shulamit Volkov, Saul Friedländer, and Dan Diner. At the Wiener Library Seminar I was fortunate to make the acquaintance of colleagues from abroad, among whom I would especially like to mention Lutz Niethammer, Andy Markowitz, Ulrich Herbert, Peter Fritsche, and Gordon Horwitz. Guli Arad and Frank Stern extended their friendship and council, while my students allowed me to try on them my more outrageous ideas and often subjected them to well-deserved criticism. As a Visiting Fellow at Princeton University I learned a great deal from Lawrence Stone, Arno Mayer, Natalie Davis, David Abraham, Martha Petrusewicz, and Sheldon Garon. Mark Mazower has kindly shared with me some of his own recent and fascinating work on the subject. I would like to thank them all. Since coming to Harvard, I have had the opportunity to exchange opinions with many more collleagues. I am grateful to Charles Maier, Stanley Hoffmann, David Landes, Jürgen Kocka, and Allan Silver, as well as to Jeffrey

Herf and Daniel Goldhagen, for having given me of their time. Paula Fredriksen and Richard Landes have shown me the extent to which rigorous scholarship, intellectual pursuits, and long-standing friendships can complement each other. I would also like to thank the readers of this manuscript for numerous helpful suggestions and comments, even if I have not always accepted their criticism. Needless to say, I alone am responsible for the final version of this work.

My research in Germany was financed by the German Historical Institute in London, the German Academic Exchange Service, and the Alexander von Humboldt Foundation. I must particularly praise the latter for its generosity and promotion of international scholarship. The Davis Center for Historical Studies at Princeton University enabled me not only to complete my previous book, but also to sketch out the first outlines of the present work. The final version of this book, however, was written at the Society of Fellows of Harvard University. I cannot sufficiently laud this institution both for enabling me to end a project which had been haunting me for years, and for providing the perfect balance between solitude and intellectual intercourse which is so hard to come by these days. It is with pleasure that I express my gratitude to the Secretary of the Society, Diana Morse, who had done more than anyone to help me achieve this peace of mind, and to all other Senior and Junior Fellows, especially Seth Schwartz, Leslie Kurke, Juliet Fleming, Robin Fleming, Chris Wood, and Moshe Halbertal. Rogers Brubaker has been a particularly demanding colleague, for while his incisive critique of my work has made me write everything twice, competing with his tremendous industry has prevented me from giving up before the finish. I can only conclude by expressing the hope that all those who have helped me along the way will not be too disappointed with the results of my efforts.

Cambridge, Mass.   O. B.
October 1990

Contents

Abbreviations

AHR	*American Historical Review*
AFS	*Armed Forces and Society*
BA-MA	Bundesarchiv-Militärarchiv in Freiburg, Germany
BAZW	*Bulletin der Arbeitskreises Zweiter Weltkrieg*
EHQ	*European History Quarterly*
EHR	*Economic History Review*
FAZ	*Frankfurter Allgemeine Zeitung*
FHS	*French Historical Studies*
FPP	Feldpostprüfstellen (Field Post Inspection Offices)
GD	Grossdeutschland Division
GFP	Geheime Feldpolizei (Secret Field Police)
GH	*German History*
GuG	*Geschichte und Gesellschaft*
H&M	*History and Memory*
Hiwis	Hilfswillige (Russian volunteers in the Wehrmacht)
HJ	Hitlerjugend (Hitler Youth)
HWJ	*History Workshop Journal*
JCH	*Journal of Contemporary History*
JfG	*Jahrbuch für Geschichte*
JMH	*Journal of Modern History*
JSS	*Journal of Strategic Studies*

LBIY	*Leo Baek Institute Yearbook*
MGM	*Militärgeschichtliche Mitteilungen*
NSDAP	Nationalsozialistische Deutsche Arbeiterpartei (National Socialist German Workers' Party)
NSFO	Nationalsozialistischer Führungsoffizier (National Socialist Leadership Officer)
OKH	Oberkommando des Heeres (High Command of the Land Forces)
OKW	Oberkommando der Wehrmacht (High Command of the Armed Forces)
P&P	*Past and Present*
POQ	*Public Opinion Quarterly*
RDP	*Revue de droit pénal militaire et droit de la guerre*
RH	*Russian History*
RHDGM	*Revue d'histoire de la deuxième guerre mondiale*
RPS	*Research in Political Sociology*
SD	Sicherheitsdienst (Secret Police of the SS)
SS	Schutzstaffel ("Defense Squad" of the Nazi Party)
TAJB	*Tel Aviver Jahrbuch für deutsche Geschichte*
VfZ	*Vierteljahrshefte für Zeitgeschichte*
W&S	*War and Society*

HITLER'S ARMY

Introduction

Almost half a century after its total destruction in the Second World War, the Wehrmacht remains a major bone of contention in the scholarship on the Third Reich. Was it merely a military organization which carried out its orders with remarkable professional skill, or a highly politicized army? Was it a haven from the regime or an exceptionally effective school of National Socialism? Did it pose a threat to Hitler's rule or was it rather his most formidable instrument? Were the generals hampered in their endeavours to topple the regime by the troops' loyalty to the Führer, or was it the army's senior officers who insisted on motivating the rank and file by large doses of National Socialist ideology? Briefly, was the Wehrmacht Hitler's army?

The following pages will argue that the only way to approach this question is by a careful anatomy of the German army. This will be done by proposing four theses on the war experience, social organization, motivation, and perception of reality of Germany's soldiers. By examining the attitudes of both the higher and the lower echelons of the army, this book will attempt to gauge the degree to which the Wehrmacht constituted an integral part of state and society in the Third Reich. Naturally, for the individuals involved things were never so neatly delineated. Yet many of the issues raised below greatly occupied contemporaries, and were not superimposed on the period merely for the sake of the argument. Moreover, differing interpretations of the Wehrmacht's position within the Nazi state have had a major impact on the postwar historiography of the Third Reich.[1] Consequently, our analysis will touch both on the "actual" historical events, and on their perception by preceding generations.

The first chapter examines the contradiction between the Wehrmacht's image as the most modern army of its time, and the profound process of demodernization it underwent particularly on the Eastern Front. By means of a detailed reconstruction of life at the front, this chapter demonstrates the effects of the immense material attrition on the troops' physical condition and state of mind. It stresses that as of winter 1941–42 the majority of Germany's soldiers were forced into trench warfare highly reminiscent of the Western Front of 1914–18, while facing, however, an increasingly modernized enemy. Unable to rely on its hitherto highly successful Blitzkrieg tactics, the Wehrmacht accepted Hitler's view that this was an all-or-nothing struggle for survival, a "war of ideologies" which demanded total spiritual commitment, and thus tried to compensate for the loss of its technological superiority by intensifying the troops' political indoctrination. This in turn opened the way for an ever greater brutalization of the soldiers.

The complex relationship between tradition, modernity, and Nazi ideology, made all the more intense by Germany's progressive material inferiority, is a fundamental problem of interpretation in the history of the Third Reich. While many senior officers upheld traditional social, political, and military values, they were attracted to Hitler not least because he made possible the rapid modernization of the army.[2] Conversely, whereas Hitler in particular manifested a great deal of fascination with modern technology, Nazi rhetoric and propaganda often expressed a powerful abhorrence of modernity and made extensive use of pseudo-religious and mythical images.[3] Initially, an attempt was made to oppose the army's professional *Soldat* to the ideologically motivated SS *Kämpfer*. However, these simplified categories tended to overlap in practice, as the former increasingly came to rely upon ideological commitment, and the latter turned out to be a highly skilled professional. A similar paradox can be observed on the strategic level. While Germany's Blitzkrieg campaigns, which were based on a rational evaluation of the relationship between economic means and military tactics, came to be considered as typically Nazi, the total war strategy, which very

efficiently mobilized Germany for the kind of war it had no hope
of winning, was seen as representative of such clear-thinking tech-
nocrats as Albert Speer.[4] Thus, instead of making the conventional
distinction between modernity and traditionalism, one might argue
that it was precisely this inherent tension between technology and
mythology, organization and ideology, calculation and fanaticism,
which constituted one of the crucial ties between the Wehrmacht,
regime, and society in the Third Reich, and provided the army with
much of its tremendous, albeit destructive energies.

The second chapter examines the destruction of the "primary
group," the social unit which had traditionally constituted the backbone of the German army. Due to the tremendous losses in the
fighting, the lack of replacements, and the rapid manpower turnover
among combat units, the Wehrmacht could no longer rely on the
"primary group" as the key for its cohesion. The widely accepted
sociological theory of Shils and Janowitz which maintains that the
Wehrmacht avoided disintegration due to its social organization is
thus shown to be largely irrelevant to conditions particularly on the
Eastern Front, where the lion's share of the army fought throughout
most of the war.[5] This makes it necessary to find another explanation
for the Wehrmacht's remarkable cohesion and battle performance,
especially in view of its aforementioned material weakness.

The thesis first propounded by Shils and Janowitz has both
influenced and reflected scholars' interpretations of collaboration
with, and dissent from, the Nazi regime in the civilian sector as
well. Indeed, one might say that there exists an as yet unrecognized
link between "primary group" theory and *Alltagsgeschichte,* that is,
between the notion that soldiers are mainly motivated by a desire
to survive coupled with loyalty to their comrades, and the increasingly popular argument that in the Third Reich most people were
far too preoccupied with everyday concerns to pay much heed to
the regime's rhetoric or policies. Hence, while "primary group"
theory "depoliticized" the Wehrmacht, one consequence of writing
the history of the Third Reich "from below" was to create an impression of a "depoliticized" civilian society, most of whose members
presumably considered the "normality" of daily life as far more
important than the "abnormality" of Nazi ideology and actions.[6]
Yet when speaking of a vast conscript army such as the Wehrmacht,
not only is it important to realize that both the soldiers' morale and

motivation, and attitudes toward the regime among civilians, were very closely related matters, but also that these are far too complex issues to be explained by means of a single, rather mechanistic and detached theory. In fact, some insight into the relationship between the people and the regime may be derived from the notion that while *real* "primary groups" do not fully explain combat motivation due to their unfortunate tendency to disintegrate just when they are most needed, the *idea* of attachment to an *ideal* "primary group," composed of a certain *category* of human beings, clearly does have a powerful integrating potential. This kind of "primary group," however, is in some respects the precise opposite of the one presented in the original theory, for it is very much the product not merely of social ties, but of ideological internalization, whereby humanity is divided into opposing groups of "us" and "them." Indeed, the sense of identification with one group, and the abhorrence of the other, are in both cases dependent on an abstraction; personal familiarity may only weaken the individual's commitment by revealing the less than ideal aspects of his own side, and the human face of his opponents (which is why armies dislike fraternization). This kind of categorization is of course just as applicable to civilians, and in both cases does not necessitate any profound understanding of whatever world-view one believes oneself to be fighting or working for. Instead, it calls for internalizing only those aspects of the regime's ideology based on previously prevalent prejudices,[7] and most needed to legitimize one's sufferings, elevate one's own status, and denigrate one's enemies, be they real or imaginary.

The third chapter proposes that it was the unprecedented harsh discipline of the Wehrmacht which kept the units together at the front. However, the soldiers' submission to a disciplinary system which led to the execution of some 15,000 men[8] was closely tied to the troops' own conduct toward enemy soldiers and civilians. While many of the army's criminal activities were directed from above, the troops went unpunished even when they totally disregarded orders forbidding plunder and indiscriminate shooting. By allowing unauthorized actions against individuals considered as mere "subhumans," the army created a convenient safety valve which made it possible to demand strict combat discipline. Cohesion came to depend on a perversion of the moral and legal basis of martial law. Nevertheless, when terror from the enemy became greater than fear of one's superiors, breakdowns did occur. Complete disintegration was prevented not merely by discipline, but by creating a commonly shared view of the war which made the prospect of defeat seem equivalent to a universal apocalypse.

Domestic "discipline" in the Third Reich, perceived by many as a return to "normality" after the chaos of the Republic, was achieved by means not unlike those employed by the Wehrmacht, that is, by exploiting the regime's popularity and the public's conformism, while simultaneously stamping out any opposition with the utmost brutality.[9] Thus, quite apart from isolated manifestations of actual social or military disobedience and revolt, our understanding of the vast majority's obedience to the commands of the regime will depend to a large degree on the relative weight attributed to willing and possibly ideologically motivated support on the one hand, and fear of punishment on the other. The severity of the Wehrmacht's discipline was not simply part of an old Prussian tradition, but rather the result of profound changes introduced into martial law under the Third Reich, as were indeed the instructions issued to the troops concerning the manner in which enemy soldiers and civilians were to be treated. The question of discipline *per se* cannot be divorced from the new ideological determinants of martial law, and any discussion of the nature of offenses and their punishment must take this factor into account if it is not totally to misinterpret the evidence. This was of course very much the case with the Third Reich's civilian society as well, for although we are speaking of two separate legal systems, it must be taken into account that both were substantially altered to fit the ideological requirements of the regime.[10] Indeed, the fact that differentiating attitudes toward various categories of human beings were neither limited to the Wehrmacht nor reserved to occupied populations could clearly be seen from the euthanasia and racial campaigns within the borders of the Reich.[11] Furthermore, the obedient and uncritical participation of millions of soldiers in "legalized" crimes was significant also in that it probably both reflected the moral values these young men had internalized before their recruitment, and affected their state of mind and conduct upon returning to civilian society following the collapse of the Third Reich. One aspect of this impact could be seen in the content and uncritical public reception of the numerous personal memoirs and formation chronicles published in Germany in the

1950s and 1960s, which revealed an alarming sympathy with the distorted norms of discipline and obedience, law and criminality which characterized the Wehrmacht.

Chapter 4 discusses the extent to which years of premilitary and army indoctrination distorted the soldiers' perception of reality. The Wehrmacht's propaganda relied on a radical demonization of the enemy and on a similarly extreme deification of the Führer. The astonishing efficacy of these images is shown by reference to a wide array of evidence, ranging from analyses of soldiers' opinions by the regime's own agencies and leaders, to the views of its opposers, the memoirs of former generals and soldiers, the oral testimonies of workers and youths, and the private correspondence of troops from the front. It is particularly in the latter case that we find how soldiers preferred to view the reality they knew best through the ideological filters of the regime. These images also played an important role in the distorted reconstruction of the memory and history of the war, as can be seen from the manner in which recent attempts to "historicize" the invasion of the Soviet Union have employed arguments lifted directly from the Nazi regime's own wartime propaganda.[12]

The most powerful and successfully disseminated argument regarding the aims and nature of the Wehrmacht's war in the East is based on relegating the issue of the army's criminal involvement to a position of secondary importance, while simultaneously placing the Wehrmacht firmly in the anti-communist camp. Indeed, this approach both strives to "balance" the barbarities of the Wehrmacht with Soviet atrocities and, even more significantly, to shift the stress to the vast service rendered by the troops of the Third Reich to Western civilization as a whole in damming the "Asiatic-Bolshevik flood." The origins of this image of the Wehrmacht as the bulwark of *Kultur* date back to Nazi Germany's invasion of the Soviet Union in summer 1941, presented at the time as a crusade against Bolshevism and achieving a certain popularity in occupied Western Europe.[13] But its greatest gains were made when the Third Reich was already in its death-throes, at a time when Nazi propaganda did its utmost to convince the troops that they were defending humanity against a demonic invasion, while simultaneously hoping to sow dissent between the Soviet Union and its Western Allies. Though not successful in preventing the total collapse of the Reich, these efforts did bear fruit in another important sense, for they both prepared the ground for the FRG's eventual alliance with the West, and provided the Wehrmacht's apologists with a forceful and politically most applicable argument, even if it conveniently confused between cause and effect.

The astonishing persistence of this new/old image of the Wehrmacht was given powerful expression in the recent *Historikerstreit*.[14] Indeed it may be that not enough attention has been focused on the bizarre inversion of the Wehrmacht's roles proposed by all three major exponents of the new revisionism, whereby overtly or by implication, the army was transformed from culprit to saviour, from an object of hatred and fear to one of empathy and pity, from victimizer to victim. Thus Michael Stürmer's geopolitical determinism adds a measure of scientific inevitability (and continuity) to Germany's historical mission to serve as a bulwark against barbarian invasions from the East; Ernst Nolte's thesis about the horrors of the Gulags and the fear of Bolshevism as having "originated" Auschwitz makes it possible to rearrange chronology and imply that the Wehrmacht's invasion of the Soviet Union was essentially a preventive attack, just as the atrocities it committed were merely intended to anticipate even worse barbarities by the "Asiatic hordes"; and Andreas Hillgruber's awe at the Wehrmacht's self-sacrificing struggle to halt the "orgy of revenge" about to be unleashed by the Bolsheviks enables him insist on the need to "empathize" with the troops of East Prussia in conscious detachment from the inmates of the death-camps whose continued extermination the *Ostheer* thus assured. Martin Broszat's much-debated "plea" for a historicization of the Third Reich, though aiming at a much more subtle kind of revisionism,[15] is therefore answered by Stürmer's and Nolte's attempts to place Nazi Germany within a larger historical context, and by Hillgruber's insistence on empathy with the individual *Landser*.

The book concludes by arguing that in Germany the popular memory of "Barbarossa" is based on the same inversion of reality which was common during the Third Reich, whereby the war's military events and physical hardships are greatly overemphasized, while its truly unique aspect, namely its inherent criminality, is repressed and "normalized."[16] Yet it is the central contention of this study that just as we cannot speak of the Wehrmacht *as an institution*

in isolation from the state, so too it is impossible to understand the
conduct, motivation, and self-perception of the *individual officers
and men* who made up the army without considering the society
and regime from which they came. And, as the relationship between
the military and civilian society was mutual rather than one-sided,
it would also be necessary to take into account the impact not merely
of the Wehrmacht establishment, but also of the millions of soldiers
who went through its ranks, upon all aspects of life in the Third
Reich.

Only by accepting this premise does it become clear that the
army was neither simply forced to obey the regime by terror and
intimidation, nor was maneuvred into collaboration by the machinations of a minority of Nazi and opportunist officers, nor, finally, supported the regime due to some misunderstanding of
what National Socialism really meant and strove for. For all these
explanations will appear insufficient once we realize that particularly and increasingly in the Third Reich, the army as an institution formed an integral part of rather than a separate entity from
the regime, while as a social organization it was composed of a
rapidly growing number of former civilians, and consequently reflected civilian society to a greater, rather than a lesser extent
than in the past. The Wehrmacht was the army of the people,
and the willing tool of the regime, more than any of its military
predecessors.

It is in this manner too that the connection recently emphasized
between the Wehrmacht's criminal conduct in the East and the ex-
tremination of the Jews should be understood, whether we speak
of the generals or the privates.[17] Indeed, although differences of age,
social background and education, political tradition, and religion all
played a part in each individual's actions, the soldiers were more,
rather than less likely than the civilians to belong to those categories
supportive of the regime, its ideology, and its policies,[18] while the
army's top echelons, with their *raison d'être* being the direction and
application of violence,[19] found it relatively easy to legitimize the
execution of Nazi policies with what seemed to be purely military
arguments. It is thus in large part the tendency to overlook or
underestimate the importance of the intimate ties between the army,
the regime, and society, rather than any "objective" lack of docu-

mentary evidence, which has hitherto made for posing the wrong questions and offering unsatisfactory interpretations as to the functions, influences, and historical importance of the Wehrmacht in the Third Reich.

1

The Demodernization
of the Front

One of the greatest paradoxes of the Second World War was that between 1941 and 1942 the Wehrmacht's combat units underwent a radical process of demodernization, just as the Third Reich's economy was being mobilized for a total industrial war.[1] The successes of the German army in the first two years of the fighting were based on an innovative and highly effective employment of its limited material resources. The Blitzkrieg campaigns in Poland, Scandinavia, and western Europe were brief and claimed relatively few casualties. Once the Wehrmacht invaded the Soviet Union, however, and particularly in the course of winter 1941–42, the realities of the front were profoundly transformed. Although the Reich actually produced a growing number of war machines, the vast majority of German combat troops lived and fought in conditions of the utmost primitiveness. As we shall see, this process had a major impact on the character, self-perception, and conduct of the Wehrmacht.[2]

When Germany launched its attack in the West, its armored forces were in fact numerically and in some respects also qualitatively inferior to those of its opponents. On 10 May 1940 the Wehrmacht sent into action 2445 of its 3505 available tanks. Facing it were no less than 3383 French, British, Belgian, and Dutch tanks. Moreover, only 725 of the German tanks were of the advanced Panzer III and

IV models, and even they had great difficulties in confronting some of the heavy French tanks. Yet the Germans made much better use of their armor, for unlike the Western Allies, who distributed their tanks piecemeal along the whole front, they organized them in Panzer divisions, and then grouped those formations into powerful fists capable of thrusting deep into the enemy's rear. These organizational innovations made it possible to achieve an overwhelming local superiority by employing entirely new tactics of concentration, breakthrough, and penetration, creating the impression of overall numerical and technological preponderance. The shock effect of such armored raids was enhanced by a similar organization and concentration of airpower. Unlike the armored corps, the *Luftwaffe* did have more and better aircraft than its opponents. On the eve of the attack the Germans had 4020 operational airplanes, as opposed to 3099 Allied machines, including aircraft stationed in Britain. Even more important, the *Luftwaffe* deployed in the West 1559 bombers, whereas the Allies had only 708, most of which were relatively obsolete. The combination of massed armored penetrations closely supported by this "flying artillery" rapidly unhinged the enemy's front, disorientated his command, made havoc of his logistical system, and greatly demoralized both front and rear. Under such circumstances the Allies' great superiority in guns proved quite meaningless. The campaign was won so swiftly and decisively that, retrospectively, both sides came to view its outcome as inevitable.[3]

In fact, victory was anything but a foregone conclusion. The Wehrmacht's armored element was merely a tiny fraction of its overall strength. The Germans attacked in the West with 141 divisions, only 10 of which were armored. In order to bring the maximum weight of tanks to bear at the *Schwerpunkt* of the battle, in the crucial stage of the campaign no less than nine Panzer divisions raced side by side to the Channel, thereby cutting the Allies' forces in two. This concentration of almost all the Wehrmacht's modern elements at one point of the front did achieve its goal. But it might have also proved fatal. Had the Allies shown just a little more understanding of the battle scene, and had they manifested a slightly greater degree of organizational skill and coordination, they could have severed this armored fist from the infantry formations trailing behind, as well as from its vital logistical support, and thus rendered it quite useless once it had exhausted its fuel and ammunition supplies. Hitler and his generals were well aware of the tremendous risk they were taking; indeed, they were haunted by the spectre of another "miracle on the Marne." But in the relatively confined spaces of the West, facing a hesitant, ill-prepared, and numerically more or less equal opponent, this was a risk well worth taking.[4] In the East things soon proved to be quite different.

The Wehrmacht achieved a tactical victory. But because Germany had not won the war, its inherent weaknesses now became increasingly evident. This was the second, and more profound lesson to be drawn from the Western campaign. The Third Reich had attempted to fight a European war without totally mobilizing its economy. Having failed to achieve total victory in the West due to Britain's insistence to continue fighting, Hitler turned eastward, hoping to destroy the Soviet Union with the same tactics which had proved so effective in the West. But between 1940 and 1941 Germany's total war production as calculated in financial expenditure had hardly grown, while that of Great Britain, the Soviet Union, and the United States put together had almost doubled, and was already three times larger than that of the Reich.[5] By winter 1941–42 Germany found itself engaged in a world war, and reluctantly shifted to total economic mobilization. But not only did it enter the race relatively late, its resources were much more limited than those of its enemies. The Third Reich was capable of winning a European Blitzkrieg; it could not win a total world war.

The risks of Blitzkrieg tactics and strategy, and the fundamental limitations of Germany's production capacity, became glaringly visible during the first six months of the Russian campaign. The Reich's war industries succeeded in raising tank production from 2235 in 1940 to 5290 in 1941; and the Wehrmacht doubled the number of its armored divisions to twenty-one (though at the price of reducing the number of tanks per division by a third). Yet this expansion of the modern combat elements proved far from sufficient in view of the tremendous losses at the front, and the size of the enemy's own armored forces. It is indeed quite revealing that judging by the ratio between manpower and fighting machines, the Soviet forces directly facing the *Ostheer* were more modern, even if just like the Western Allies, they too had not yet learned to make effective use of their material strength. In June 1941 the *Ostheer*'s 3,600,000 troops attacked with 3648 tanks out of a total German stock of 5694;

once again, only 444 of these were of the relatively advanced Panzer IV model. Facing it in Western Russia were 2,900,000 Soviet soldiers supported by no less than 15,000 tanks out of a total armored force of 24,000, more than all the tanks in the rest of the world put together. To be sure, the vast majority of Soviet machines were quite obsolete, but 1861 were T-34 and heavy KV tanks, significantly superior to the best machines produced at the time in Germany. And, whereas in 1940 only 358 such tanks were built in the Soviet Union, in the first half of 1941 alone their number rose to 1503, and even in the second half of that year, in spite of the occupation of Russia's primary industrial regions, a further 4740 advanced models were turned out. Similarly, the *Ostheer* was supported by only 2510 aircraft, considerably less than it had deployed in the West, whereas the Russians had up to 9000, though in this case they were generally inferior to German planes. Worse still, once the Blitzkrieg faltered, the Soviet Union's greater manpower and industrial resources came into play and rapidly widened the technological margin between the Red Army and the *Ostheer*.[6]

It should again be stressed that following the debacle of winter 1941–42, the Third Reich both immensely expanded its overall war production and made considerable strides in the development of some highly sophisticated weapons and machines. Yet the experience of the average combat soldier at the front did not reflect Germany's switch to total war production. This was so both because proportionately the enemy became ever stronger and better equipped, and because due to the vast expanse of the territories occupied by the Wehrmacht, the Reich's rising production figures seemed far less impressive at the receiving end. In the relatively constricted spaces of the West, the Wehrmacht's policy of maintaining a few well-equipped divisions at the expense of the great bulk of infantry formations had proved effective. In the East, one of the keys to the failure of the Blitzkrieg was the infantry's inability to keep up with the armored spearheads over a long distance. Consequently the nature of the war changed drastically as a more or less stable front emerged which could only be held by the Wehrmacht's ill-equipped infantry formations, along with a growing number of armored divisions which had lost most of their tanks. Only a few elite units were kept well supplied with modern fighting machines, but they were no longer able fundamentally to change the overall situation.

This was the reason that although as viewed in overall production figures the Wehrmacht seemed to be undergoing a process of modernization, the experience of most troops on the ground was of profound demodernization, of a return to the trench warfare of the Great War made worse by the enemy's growing technological capacity.

It is of some interest to examine these developments at greater detail. As far as war production was concerned, Germany managed to raise the annual turnout of light and medium tanks to as many as 22,110 by 1944, at which time it was also producing 5235 superheavy tanks. Yet the Soviet Union maintained an annual production rate of 30,000 tanks as of 1943; Britain produced 36,720 tanks between late 1941 and the end of 1943; and the United States turned out a total of 88,410 tanks. Similarly, the Third Reich raised aircraft production from 12,401 in 1941 to 40,593 in 1944. But the Soviet Union achieved a monthly production rate of between 2000 and 3000 aircraft in the last years of the war, and the United States produced just under 100,000 fighter planes and more than 90,000 bombers, of which more than a third were four-engined long range machines. All this was apart from the tremendous output of the American motor industry, which turned out well over four million armored, combat, and supply vehicles of all kinds, a large proportion of which were instrumental in the motorization of the Red Army.[7]

From the perspective of the front, the length of the Soviet frontier meant that the *Ostheer* could repeat its Blitzkrieg tactics only by splitting its armor among three army groups, making each of its spearheads weaker than the single and decisive armored concentration of the Western campaign. In the central sector of the front, where the most powerful grouping of German forces was to be found, it was necessary once more to split the armored formations between two Panzer groups in order to encircle the large Soviet forces in Belorussia. As the Germans charged deeper into Russia, the length of the front almost doubled, from 800 to 1500 miles, and lines of supply extended some 1000 miles to the rear. This further increased the dispersal of tank units, and caused tremendous difficulties with maintaining their crucial logistical links to the depots. Matters were made worse due to the fact that while the *Ostheer's* supply apparatus was inadequately motorized, unlike western Europe, in Russia roads were sparse and mostly unpaved, and the railroad used a different gauge. It is indicative of the partial modernization of the Wehrmacht that although the Panzer divisions were given their own motor supply columns, fully 77 infantry divisions, or about half of the entire invasion force, relied strictly on horse-drawn wagons for their provisions from the railheads. Moreover, lack of spare parts meant that damaged vehicles could not be repaired, while overexertion and lack of food greatly diminished the number of horses. Because the northern and southern army groups failed to reach their operational goals with their own limited tank forces, armored units in the center had to come to their assistance, thereby greatly weakening the *Ostheer*'s thrust toward Moscow. By the time the Germans finally made up their minds to attack the Russian capital, their material strength had already significantly diminished, and their logistical system was in increasing disarray. In retrospect it can be said that the attempt to repeat the risky Blitzkrieg tactics employed in the West with an even less favorable ratio between space and machines was thus bound to fail.[8]

Matériel and mentality were closely related matters in the *Ostheer*. As the number of the tanks diminished, the troops had to dig in and revert to trench warfare; as the trucks broke down and the trains failed to arrive, provisions of ammunition, food, and clothes decreased. The demodernization of the front was thus a process whereby the disappearance of the machine forced the individual soldier into living conditions of the utmost primitiveness. The nature of this process can be seen from the following instances. Panzer Group 4, the armored element of Army Group North, raced over 200 miles into Soviet territory in the first five days of the campaign, but then had to wait a whole week for its supply basis to be pushed forward before it could renew its advance. Even then, it was necessary to divert all army group provisions and transport resources to the the tank units, with the result that the infantry was left far behind. Hence when the armored divisions reached the gates of Leningrad, they had to wait so long for the infantry to catch up with them that in the meanwhile the city's defenses were reinforced and it was no longer possible to capture it. From this point on the front solidified and the remaining tanks were diverted to the center.[9] Similarly, further south the infantry soldiers of 16 Army marched on foot over 600 miles during the first five weeks of the campaign, and then found themselves in an area of swamps east of the Lovat

River where they were to remain for the next fourteen months in the most wretched conditions conceivable.[10] The exertions of infantry troops in the initial stages of "Barbarossa" were not unusual, albeit especially great, for an army which had conducted all its Blitzkrieg campaigns with a small motorized element supported by a vast mass of walking infantry. But when the tanks got bogged down, and the whole army became stranded deep inside the Soviet Union, the infantry became the backbone of the front, clinging to it with the same desperation as the men of 1914–18, and with just as little hope of ever being rescued from their predicament by this or that technological means. Throughout the front lack of fighting machines combined with the climatic and geographical peculiarities of Russia to deprive this former Blitzkrieg army of all semblance of modernity. The commander of 16 Army's II Corps reported as early as 28 October 1941 that

> The recent rainy weather has made the roads and terrain . . . so impassable, that only tractors, [Russian horse-drawn] Panje-wagons and cavalry can still maintain a limited degree of mobility. . . . From my own experience I know that while walking on the roads one sinks to one's knees in the mud, and the water pours into one's boots from above. Fox-holes are collapsing. . . . Some of the troops have been eating only cold food for many days, as the field kitchens and Panje-wagons could not get through and the number of food-carriers did not suffice.

It is no wonder that these conditions had a direct impact on the soldiers' physical health and state of mind. As the corps commander added,

> The health of men and horses is deteriorating due to the wretched housing facilities. . . . The men have been lying for weeks in the rain and stand in knee-deep mud. It is impossible to change wet clothing. I have seen the soldiers and spoken with them. They are hollow-eyed, pale, many of them are ill. Frostbite incidence is high.[11]

The situation of the German infantry formations south of Leningrad became much worse when six divisions were encircled by the Red Army in the area of Demyansk. Between February and April 1942 these 96,000 men found themselves in an extremely precarious operational and logistical predicament. Here again the failure of technology resulted in tremendous misery. As the *Luftwaffe*'s promises to airlift supplies to the besieged troops were fulfilled only to a very limited degree, the men were compelled somehow to fend for themselves with ever diminishing provisions of food, wearing tattered summer uniforms with hardly any shelter from the fierce weather, and fighting an increasingly well-equipped enemy with ineffective weapons and never enough ammunition.[12] The demodernization of the Eastern Front was nowhere more evident than in the Demyansk "pocket": soldiers insulated themselves from the cold with newspapers, until those too ran out; boots, gloves, caps, sweaters, and coats were all in very short supply, and hardly appropriate for Russian winter conditions in any case; the meager food rations invariably reached the front-line frozen and consequently were hardly edible.[13] The 12th Infantry Division's doctor reported that the troops were living in dark, damp bunkers, badly ventilated and cramped, almost impossible to heat and thus offering little opportunity to rest from action out in the open. The foul air caused numerous respiratory diseases, while due to the absence of any means for washing and cleaning clothes, the men were all infested with lice and suffered from endless skin infections. The extreme cold and indifferent hygiene also led to frequent inflammations of the bladder and a high incidence of frostbite. Manpower shortage meant long hours of guard duty, and the consequent lack of sleep along with the incessant tension had a debilitating effect on the troops, making them, in the doctor's words, increasingly apathetic (*"geistig immer Stumpfer"*).[14] There were reports of soldiers fainting from exhaustion while on guard duty, as well as of cases of nervous disorders. One regimental doctor summed up the situation in the following manner:

> The men are greatly overstrung. This is becoming more visible every day, in loss of strength, loss of weight, and increasing nervousness, and has a progressively negative effect on battle performance with the accompanying appearance of friction, breakdowns, and failures on the part of commanders and men as a result of over-fatigue and overstrain of the nerves.[15]

Meanwhile, the only means left to the troops to defend themselves from Soviet tank assaults were small caliber anti-tank guns. Commanders tried to persuade the men that the "fact that shells bounce off the tanks should not serve as evidence that our guns cannot penetrate the Soviet tanks," and that "operating our *light anti-tank gun* courageously has brought good results,"[16] but this was obviously of small comfort to men who only a few months earlier were marching behind their own seemingly invincible Panzers.

The situation on other sectors of the *Ostfront* was essentially the same. Less than a month after "Barbarossa" was launched, Army Group South had to replace half of its trucks with Russian horse-drawn "Panje-wagons" due to mechanical failure and lack of replacements, and by November its armored component, now rather absurdly renamed 1 Panzer Army, had lost so many tanks and trucks that it declared itself "incapable of conducting mobile warfare." In fact, as early as September almost two-thirds of the *Ostheer*'s tanks were out of action. By mid-November Panzer Group 2, which had set out as one of Army Group Center's two armored concentrations with 1000 tanks and had received another 150 as reinforcements, was reduced to a mere 150 tanks, while only 15 percent of its supply vehicles were still functioning.[17] A look at one of this Panzer group's formations will illustrate how this tremendous attrition was experienced by front-line tank units. The 18th Panzer Division began the campaign with over 200 tanks, but due not least to costly encounters with superior Soviet T-34s, after two weeks of combat it was left with only 83 machines, or under 40 percent of its initial force.[18] Even at this early stage the divisional commander felt it necessary to warn: "This situation and its consequences will become unbearable in the future, if we do not want to be destroyed by winning [*wenn wir uns nicht totsiegen wollen*]."[19] But by 24 July, after a month of fighting, the 18th Panzer was left with only twelve tanks,[20] which meant that it could no longer be considered an armored formation. Nor was the division's logistical situation much better, especially after the supply column of the Panzer regiment was wiped out in a Soviet tank raid.[21] This material destruction was accompanied by manifestations of extreme battle fatigue, caused by a combination of inhospitable terrain, tenacious enemy resistance, and the Red Army's preponderance in tanks and artillery. As early as mid-July, when the Blitzkrieg was still in full swing, the 18th Panzer's motorcycle battalion experienced ten days of defensive trench warfare under constant Soviet artillery barrages and infantry assaults, highly reminiscent of the Great War. This was one of the first instances of what only a few months later was to become

the norm on the Eastern Front, namely, that material attrition on
the German side, and increasing material strength on the Russian
side, forced even many of the *Ostheer*'s armored units to dig in and
fight for their lives. In this case the result was group battle fatigue,
forcefully described by the battalion doctor:

> A state of absolute exhaustion is noticeable . . . among all men of
> the battalion. The reason is . . . a far too great mental and nervous
> strain. The troops were under a powerful barrage of heavy artillery.
> . . . The enemy charged them . . . penetrated their positions and was
> repulsed in hand-to-hand fighting . . . the men could not shut their
> eyes day and night. Food could be supplied only during the few
> hours of darkness. A large number of men, still serving with the
> troops at present, were buried alive by artillery fire. That the men
> were promised a few days of rest . . . but instead found themselves
> in an even worse situation . . . had a particularly grave effect. The
> men are indifferent and apathetic, are partly suffering from crying
> fits, and are not to be cheered up by this or that phrase. Food is
> being taken only in disproportionately small quantities.[22]

As an important element in the *Ostheer*'s campaign, the 18th
Panzer was reinforced, though typically it remained far weaker than
it had been on 22 June, and in mid-August numbered less than fifty
operational tanks, or merely a quarter of its initial force.[23] Moreover,
due to lack of motorized transport the division now had to rely on
a newly established Panje-wagon supply column, certainly not the
most appropriate means of provisioning an armored unit highly
dependent on speed and maneuvrability.[24] As the fighting dragged
on and no end was in sight, more and more troops became aware
that the war was taking on a new character. One soldier diagnosed
the cause of the Wehrmacht's approaching failure in Russia with
remarkable precision and foresight:

> Today three months ago the campaign against Russia began. Every-
> body supposed at the time that the Bolsheviks would be ripe for
> capitulation within no more than eight to ten weeks. . . . That as-
> sumption, however, was based on a widespread ignorance of the
> Russian war matériel. . . . We were spoilt by the preceding
> Blitzkriege.[25]

Indeed, when the last phase of the campaign began with the attack
on Moscow, the remnants of the *Ostheer*'s modern elements were

rapidly destroyed. The 18th Panzer was reduced to only fourteen tanks by 9 November, and ten days later whatever had remained of its armor was put out of action due to lack of fuel.[26] On the eve of the Soviet counter-offensive in early December, the division had only a quarter of its original mobility.[27] This in turn also meant severe shortages in supplies of food and clothes. The harsh living conditions naturally made for a great rise in the incidence of frost-bite, disease, and exhaustion.[28] The retreat from the attacking Red Army made things even worse, for much of what had remained of the division's matériel had to be left behind.[29] Symptoms of mental attrition caused by fatigue, hunger, exposure, and anxiety also became increasingly prevalent. On 22 December the division's Operations Section noted: "The physical and mental condition of the soldiers and of some of the commanders calls for issuing very detailed orders and carefully examining them, in order to avoid breakdowns."[30] Indeed, on Christmas eve two young soldiers died of exhaustion.[31] Apathy was widespread. A soldier from the 57th Infantry Division wrote at the time that due to the "snow-storms, blizzards and the great cold reaching down to 45 degrees [Celsius] ... there are many men who cannot find enough energy to withstand the severity of winter and escape an otherwise certain death."[32] The demodernization of armored formations did not consist merely in losing their tanks and trucks, but just as much in the numerous cases of physical and psychological breakdown caused by the wretched living conditions. As the divisional Quartermaster noted, "the almost inhuman strain" of combat was made

> even worse as there were no accommodation facilities. In spite of the great cold, reaching 40 degrees [Celsius] below zero, the troops had to spend day and night in the open and were only beginning to dig fox-holes and set up accommodation facilities . . . [when] the division received marching orders to a new assignment.[33]

Throughout the rest of the first winter in Russia the 18th Panzer's troops remained in essentially the same conditions as those of the 12th Infantry. This was characteristic of most *Ostheer* armored divisions which, stripped bare of their matériel and forced into entrenched, defensive warfare, underwent a relatively far more radical process of demodernization than the regular infantry. Later on the 18th Panzer did manage to scrape together a few tanks, but their number never rose much above twenty, and was usually far lower than that. In fact, the division's material condition was even worse than that of most Great War formations, for during the second part of winter 1941–42 it never had more than thirty-two guns, and often as few as five, while due to a great shortage of motorized supply vehicles and the high mortality of horses, it was quite immobile for much of the time, relying on primitive Russian Panje-wagons and sledges for its provisions. Another indication of the decay of the front was the quality of the equipment. While even the best German tanks were inferior to their Russian equivalents, by now Panzer units were compelled to fall back on more obsolete models, quite useless against Soviet armor; similarly, the few artillery pieces left were so worn out that their barrels were in imminent danger of exploding, while anti-tank guns, as was the case in the 12th Infantry, were mostly ineffective due to their small caliber.[34] The sense among the troops that they had been plunged into a state of extreme backwardness was enhanced by the fact that although this division, rather exceptionally, had a relatively large though still insufficient stock of winter clothes, these were far from adequate for Russian winter conditions.[35] And, although front-line units kept reporting this to the rear, the quality of Wehrmacht winter uniforms showed little improvement well into 1943.[36] Nor did it occur to those in charge of tank production that equipping them with heating systems might be essential for fighting in the East.[37]

All these factors combined to cause the same physical and mental attrition among the troops of the 18th Panzer which we noted in the 12th Infantry. Badly fed and clothed, filthy and infested with lice, lacking shelter and fighting against increasing enemy pressure with diminishing manpower reserves, the troops were assailed by an array of diseases, ranging from influenza, skin infections, and frostbite, to intestinal inflammations and typhus and spotted fever epidemics.[38] In mid-February 1942 the 18th Panzer's commander pointed out that

> due to the constant great demand of guard duty and patrols, and furthermore because of the wretched accommodation facilities, a significant deterioration in the physical and mental resistance strength [of the troops] can be observed . . . one company was pulled out of the line and quarantined owing to frequent outbreaks of spotted fever. . . . The reduction of food rations is unbearable in the

long run in view of the troops' condition. A full rest is advisable for replenishment, restoration of health and morale.[39]

As rest was out of the question, it is not surprising that during the first three months of 1942 some 5000 soldiers, or close to a third of the division's manpower, reported sick.[40] This was of course by no means exceptional. In 1942 spotted fever reached epidemic proportions throughout the *Ostheer*, claiming no less than 36,434 victims. Indeed, in December 1941 alone the number of the sick in the East rose steeply to 90,000, along with a similarly high incidence of frostbite. By January 1942 almost two-thirds of the overall 214,000 troops lost to the *Ostheer* were victims of illness and frostbite rather than enemy action, with their numbers rising to as many as half a million by spring 1942.[41] The wretched living conditions and the resulting deterioration in the troops' health also had a serious psychological effect. The commander of the 18th Panzer was not alone in urging his officers "to take vigorous action against the clearly widespread fatigue and indifference among our men."[42] The first winter in Russia both materially demodernized the German side of the front and produced a changed mentality among the Wehrmacht's troops. Symptomatic of this metamorphosis was the complaint made by the commander of Army Group Center, Field Marshal von Bock, that "our troops run away whenever a Russian tank appears," an unprecedented phenomenon in an army which had only recently introduced modern armored warfare to Europe. The man who had led this military revolution, General Heinz Guderian, now described his once invincible Panzer group as "a lot of armed camp-followers who trudge slowly backwards," plagued by "a serious crisis of confidence" prevalent both "among the troops and the junior commanders."[43]

Yet the changing character of warfare on the Eastern Front did not merely produce fatigue and apathy, but also a new image of war. Between 1918 and 1939 men had come to accept the idea that the combination of the machine-gun and barbed wire which had caused the stalemate along the continuous Western Front of the Great War, would remain a permanent feature of modern warfare.[44] But during the first two years of the Second World War the Wehrmacht's tanks and aircraft accelerated the pace of fighting to such a degree that the notion of a front seemed to have vanished altogether.

The image of war became one of highly trained professionals wielding sophisticated fighting machines and conducting rapid and spectacular campaigns. Once "Barbarossa" was launched, however, the striking imbalance between the *Ostheer*'s modern and obsolete elements, already evident in the Western campaign, could no longer be bridged due to the much greater spaces of Russia and the strength and determination of the Red Army. As a consequence, the *Ostheer* became stranded along an overextended front deep inside the Soviet Union. The front assumed once more the character associated with the Great War, and the *Ostkämpfer*'s war came to consist of traditional, entrenched warfare, rather than of the quick marches and decisive encounters he had learned to expect. "To say: 'Even a dog wouldn't go on living like this,' means little, for hardly any animal lives in lower and more primitive conditions than us," wrote one soldier as early as 18 August 1941.[45] But his was a much more frustrating and demoralizing situation than that of his Great War predecessors, for unlike them he was well aware of the now lost potential of conducting war quickly, conclusively, and at a relatively low cost. Worse still, now his enemies began increasingly to inflict upon him what only a year earlier he had inflicted on them, using those technologies and tactics he could no longer employ. As one soldier wrote home, "I didn't know what trench warfare was like, but now I have learned. Our casualties are great, more than in France." But, as he went on to say, this was not simply a repetition of 1914–18, for now he was made to feel as the French and Poles had felt: "I have never seen such tough dogs as the Russians, and it is impossible to tell their tactics in advance, and above all their endless matériel, tanks and so forth."[46] This could indeed read as a letter written by a Frenchman in May 1940. Thus for the individual *Landser*, technology was transformed from an ally to an enemy, and his animosity toward its deadly products hurled at him at the front was only further enhanced by the growing intensity and ferocity of the Western Allies' strategic bombing of civilian centers and industrial targets far in the rear. The old romantic view of war was badly shaken. One soldier complained: "Yet why is this suffering in itself not great, but so unspeakably common and dirty?"[47] Instead, combat troops now began to rationalize these developments by employing a sort of nihilistic, social-darwinian argumentation, not unlike that of Ernst Jünger, according to which not only was

war hell, one also had to be a beast if one wished to survive it. The
chronicler of the elite Grossdeutschland Division, who had served
in its ranks as an officer, succinctly described this new attitude:

> Man becomes an animal. He must destroy, in order to live. There
> is nothing heroic on this battlefield. . . . The battle returns here to
> its most primeval, animal-like form; whoever does not see well,
> fires too slowly, fails to hear the crawling on the ground in front
> of him as the enemy approaches, he will be sent under. . . . The
> battle here is no assault with "hurrah" cries over a field of flowers.[48]

Another soldier wrote as early as 11 July: "Here war is pursued in
its 'pure form' ['Reinkultur'], any sign of humanity seems to have
disappeared from deeds, hearts, and minds. The scenes which one
observes border on insane hallucinations and nightmares."[49] And
Ansgar Bollweg, a thirty-year old student of theology, mentioned
Jünger approvingly when he summed up his own view of the war
in a letter written in November 1943: "This war has made us soldiers
different. . . . With the senses of a predator we recognize how the
rest of the world will be ground between the millstones of this war.
The Middle Ages are finally reaching their end. Knights, kings,
townsmen, peasants have all been destroyed."[50] Paradoxically, the
troops may have clung to each other and kept on fighting precisely
because of that terrible sense of isolation and abandonment which
oppressed them so heavily, for there was nowhere to flee to in the
depths of Russia:

> It is the courage of the desperate [writes the chronicler of the GD],
> trying to defend what has already been won, the fear of falling alive
> into the hands of the enemy, and the instinct of self-preservation,
> which are the reasons for the willingness of the men fighting in the
> East to make this sacrifice. They do not give up.[51]

This was indeed a new concept of heroism, a new self-perception
of the combat soldier, which substituted a ruthless, fanatic, amoral
view of war for material strength and rational planning. There was
an anarchic element in this celebration of death and return to sav-
agery among the front-line troops, combining a growing contempt
for traditional authority and values with a powerful urge to anni-
hilate both one's enemies and oneself. The origins of this view, not
unrelated to the romantic imagery of war, preceded the Russian

campaign. Even in 1939 the twenty-year old Baron von Guttenberg wrote from Poland: "War is a blood-letting to be endured by humanity. It is the duty of every man to conduct himself in harmony within the chaotic confrontation of storming spirits." And the young Oskar, Prince of Prussia, rejoiced at the time: "We stand before the burning gates of Europe . . . and only a shudder of belief illuminates our actions, of which one says he would wish to follow us to the end of the world."[52] Another soldier wrote from France in June 1940 that "war is and will remain a condition of existence. A state, a national community, appears to need periods of fighting, in order to preserve its values and to fulfil its tasks; otherwise it must surrender them by becoming powerless and weak."[53] Yet the unprecedented turmoil on the Eastern Front greatly radicalized this concept, producing the image of the new, ideal, instinctive warrior. "Orders are not given any more," wrote one member of the GD Division: "Leadership has reverted to its original form." It is, he claimed, "a battle for survival," where everything was allowed which might prevent the extinction of the individual soldier, and by extension also that of his comrades, his unit, his race and country.[54] But it became much more than that, for in the attempt to overcome the material demodernization of the front, the terrible mental attrition of the soldiers, the hopelessness of the situation, and the ever growing superiority of the enemy, battle was made into a thing of itself, a condition to be glorified as the real, supreme essence of being. Indeed, even those who perceived the madness of the war saw no other way of coping with its reality than by idealizing it. Harald Henry, a twenty-two-year-old student of philosophy, maintained in October 1941:

> Yes, what I live here *is* Idealism. The Idealism of "in spite of it all," [living] right on the edge. When I make front against "idealistic" conceptions, then this is due to my bitterness against all false affirmations, against an enthusiasm that recognizes and knows nothing of what we suffer and what is being destroyed here. . . . But what we must do here, suffering to insanity, holding on and extricating ourselves with clenched teeth, ever participating and prepared—and then in the midst of the atrocious wretchedness, in the abysses and darkest aspects of life, still preserving the belief in its light and beautiful sides, in the meaning of life, in the external values, in the whole rich and beautiful world of Idealism, how

should we call that? This is that "in spite of it all," that inner indestructibility, that unshakable will to conceptualize even the most horrible as part of the whole, to see it within the "good" general stream of life.[55]

The demodernization of the front had several important consequences. First, it led to such heavy losses among combat units that the traditional backbone of the German army, the "primary groups" which had hitherto assured its cohesion, were largely wiped out. Second, in order to prevent the disintegration of the army as a whole which might have resulted from the breakup of the "primary group," the Wehrmacht introduced and ruthlessly implemented an extremely harsh disciplinary system, to which was given not merely a military, but also an ideological legitimation. Yet draconian punishment did not suffice in cases where fear of the enemy was greater than fear of one's superiors. Thus in compensation for their obedience, and as a logical conclusion of the politicization of discipline, the troops were in turn given license to vent their anger and frustration on the enemy's soldiers and civilians. The demodernization of the front consequently greatly enhanced the brutalization of the troops, and made the soldiers more receptive to ideological indoctrination and more willing to implement the policies it advocated. This process was possible, however, only because a large proportion of the Wehrmacht's officers and men already shared some key elements of the National Socialist world-view. Confronted with a battlefield reality which no longer corresponded to their previous image of war, and with an enemy who could not be overcome by employing familiar military methods, German soldiers now accepted the Nazi vision of war as the only one applicable to their situation. It was at this point that the Wehrmacht finally became Hitler's army.

2

The Destruction of the Primary Group

The most obvious, and to contemporary German soldiers also the most painful consequence of the failure of the Blitzkrieg in the Soviet Union and the demodernization of the front was the tremendous casualties sustained by combat units. This was neither a local nor a temporary setback, for from the very beginning of "Barbarossa" until the final stages of the war the Wehrmacht invested the lion's share of its manpower and material resources in the East, and as of winter 1941–42 was never again able to seize the initiative on more than an increasingly limited part of the front. Most *Ostheer* soldiers spent the next three years of the war in defensive positions and predictably sustained the kind of losses associated with entrenched warfare which previous Blitzkrieg campaigns had managed to avoid. Indeed, it was in the Soviet Union that the Wehrmacht's back was broken long before the Western Allies landed in France, and even after June 1944 it was in the East that the Germans continued to commit and lose far more men. Hence, the war experience of most German combat soldiers was forged on the Eastern Front, and it is only by examining events there that one can gain the proper insight into the functioning of the Wehrmacht, the mentality and self-perception of its troops, and the changes it underwent during the war. Put differently, one can argue that only during the war in the

Soviet Union did the Wehrmacht attain full "maturity" and finally become Hitler's army. This is important to recognize not merely because of the numerical weight of the *Ostheer,* but also because it revealed the inherent dynamic of serving as the military instrument of the Nazi regime. Ultimately, not even the remotest and most isolated elements of the German armed forces could remain completely immune to this process.

Cohesion in the German army was to a large extent maintained by a conscious and systematic nurturing of what has come to be termed "primary groups." The roots of this social organization go back to a military tradition which expected soldiers to feel a special kind of bond and loyalty to their unit. This sort of *esprit de corps* was especially effective in cases where battalions, regiments, and later on also divisions were raised on a regional basis, making for linguistic, religious, normative, and many other kinds of affinities between the men. As armies expanded, this type of unit loyalty became more difficult to maintain. While the size of military formations made familiarity between all members impossible, the growth of the nation-state was accompanied by the attempt to stress national patriotism, rather than regional loyalty, as a major factor in the cohesion and motivation of the soldiers. The effort to nurture personal ties between soldiers also seemed to preclude full exploitation of the nation's human resources. Some countries, such as the United States, chose a more efficient manpower policy which treated soldiers as individuals to be sent wherever needed, rather than to a specific unit where they might have already formed, or were expected to develop personal ties. The German army, however, chose more systematically to organize the tradition of keeping together soldiers belonging to a specific unit and recruited originally from the same *Wehrkreise,* or conscription zones. This policy meant that not only were troops trained and grouped into units together, but also that the wounded could expect to rejoin their old comrades once they recovered. This practice could become administratively very cumbersome, but was highly conducive to morale, for German soldiers could see their unit as a kind of home to which they could always return, a social group made of men they knew and trusted. While it was no longer possible to maintain group loyalty to a whole regionally conscripted division, within that formation men forged "primary group" ties at the company and platoon level; to a large extent, they developed the same ties with their junior and middle-ranking officers at the platoon, company, battalion, and often also the regimental level. Indeed, the German army traditionally expected its officers not only to lead their men into battle, but also to care for their needs, creating thereby a sense of belonging to a family, albeit a highly hierarchical and disciplined one, reflected in the customary junior commander's form of address to his men as *Kinder*.[1]

This traditional social organization at the lower echelons made such a profound impression on some Western scholars studying the German army during and immediately after the Second World War that it was proclaimed as the real cause for the Wehrmacht's remarkable cohesion, rather than any ideological motivation. The gist of the argument was that

> the unity of the German Army was in fact sustained only to a very slight degree by the National Socialist political convictions of its members, and that more importantly in the motivation of the determined resistance of the German soldier was the steady satisfaction of certain *primary* personality demands afforded by the social organization of the army.[2]

The cohesion of the Wehrmacht was thus said to have been the product not of abstract ideas, but of a concrete and clearly identifiable social system which catered to the formation and preservation of close personal ties between soldiers within a network of "primary groups." As these "primary groups" constituted the backbone of the army, it was further maintained that

> once disruption of primary group life resulted through separation, breaks in communications, loss of leadership, depletion of personnel, or major and prolonged breaks in the supply of food and medical care, such an ascendancy of preoccupation with physical survival developed that there was little "last ditch" resistance.[3]

Ever since it was first formulated in 1948, this theory has had a major impact on all subsequent writing on the Wehrmacht, as well as on soldiers in general.[4] The idea that German soldiers were motivated by organization rather than indoctrination became so predominant, that very little effort was made to re-examine the evidence on which the original thesis had been based. Instead, forty years after Shils and Janowitz had suggested their "primary group" theory, another scholar similarly asserted that what he termed the outstanding "fighting power" of the German army, "rested almost solely on the excellence of its organization per se."[5] Unfavorably comparing the U.S. Army with the Wehrmacht, he too asserted that

> the average German soldier . . . did not as a rule fight out of a belief in Nazi ideology. . . . Instead, he fought for the reasons that men have always fought: because he felt himself a member of a well-integrated, well-led team whose structure, administration, and functioning were perceived to be, on the whole . . . equitable and just.[6]

We shall have an opportunity to ask what the implication is of averring that German soldiers believed themselves members of an institution based on justice and equality, and what impact this self-perception might have had on the troops' conduct. In the present context the main point to be made is that the evidence presented in support of these assertions is, to say the least, highly problematic. Shils and Janowitz had collected the opinions, through interviews and questionnaires, of German prisoners of war captured by the Western Allies during the last phase of the war and immediately after the cessation of hostilities. A number of factors cast doubt on the validity of these soldiers' testimonies and on the extent to which they were representative of the "average" German soldier. First, these men were captured while fighting against the Western Allies, whereas the vast majority of Wehrmacht troops were deployed in the East; even if some of the soldiers in question had probably been on the Eastern Front earlier on, their self-perception and attitudes to the enemy and the regime must have been molded by their recent experiences in the West. Second, considering their status as prisoners, these men could hardly be expected to reply sincerely to questions posed by their interrogators regarding their commitment to a regime and ideology deemed criminal by the enemy; moreover, POWs would generally tend to be more critical of their military and political leadership than the average soldier, for the very fact of their falling into the hands of the enemy may either undermine their confidence in their leaders or reflect a previous state of demoralization which had led to their giving themselves up in the first place. Third, these men were interviewed when the Third Reich was on the verge of collapse, or had already capitulated following Hitler's suicide, and consequently their psychological state must have been very different from that of POWs captured even a few months earlier, let alone of soldiers still fighting with their units. Finally, it is by now commonplace among historians that oral testimonies, even when taken under far less problematic circumstances, must be examined against more objective types of evidence before their historical value can be determined, as has indeed been shown by scholars concerned both with the Third Reich and numerous other societies.[7]

Even more disturbing than these reservations is the fact that "primary group" theory actually reversed the chronology of events. The Wehrmacht began to manifest its most remarkable "fighting power" precisely at a time when the network of "primary groups" which had ensured its cohesion during previous Blitzkrieg campaigns began to disintegrate, and its troops went on fighting with astonishing determination throughout the last three years of the war in spite of the fact that this supposedly crucial element of *esprit de corps* had been substantially weakened. Ideological motivation was not indispensable, though it was far from absent, during the rapid victories in Poland and France; there the tactics discussed in the previous chapter and the tightly knit social organization of the units were indeed quite sufficient to maintain the cohesion and morale of the troops. But during the first six months of fighting in the Soviet Union most of the preconditions presented by Shils and Janowitz as bound to lead not only to the disintegration of the "primary group," but by extension also to the breakup of the army as a whole, already existed. Yet while the "primary groups" did more or less disappear, the army fought with far greater determination and against far greater odds than at any other time in the past. We have already seen the material demodernization on the Eastern Front, where most of the heavy fighting was conducted throughout these years, as well as the ensuing mental attrition of the troops, both of which reached such an extent as to have warranted a complete disintegration of the army according to "primary group" theory. In the following pages we shall see in greater detail the degree to which "primary groups" by and large ceased to exist among combat troops due to the enormous casualties and rapid manpower turnover at the front.[8]

"Primary group" theory maintains that social organization

made ideology unnecessary. This is a far too dichotomous view even under the most favorable circumstances. Indeed, German soldiers had no trouble noting the importance of both. Karl Fuchs wrote from Russia on 26 October 1941, even before the massive losses of the winter months decimated his unit, that "the secret behind our incredible successes and victories" was to be found in the "great comradeship [that] binds us German soldiers together" *and* in their "devotion to the cause."[9] Egon Freitag, a twenty-three-year old student of engineering, insisted on 28 August 1941: "We were never mercenaries, but—to use the hackneyed phrase—defenders of the Fatherland. There are certainly among our ranks those who fight for the idea of National Socialism, and others who fight for the Fatherland, that spot on the map for which risking one's life remains self-evident. We lie together in the tent."[10] For him, comradeship and ideology (whether Nazi or nationalist) were as clearly inseparable as Germany's invasion of the Soviet Union was a defensive operation. To be sure, individual motivation had many complex sources. Helmut von Harnack, a twenty-three-year-old high school graduate who returned to the front after being wounded twice, wondered in a letter dated 23 September 1941 whether he was motivated by "pride and impatience," by the "sense of duty that one must help the comrades stuck in the mud, that one simply belongs there, that one cannot be torn away, because one feels oneself out there almost at home," or perhaps, citing Rilke, that he kept going merely *"um wiederzukehren!"*[11] As casualties on the Eastern Front mounted, however, the weakening of the previously effective "primary group" ties increasingly tilted the balance in favor of ideological motivation. Indeed, the fact that in the West there was relatively little "last-ditch" resistance proved precisely the opposite of what Shils and Janowitz have tried to demonstrate. In the West, due to the relatively less costly fighting, "primary groups" survived to a much larger extent than in the East. The reason that the *Westheer* presented generally little "last-ditch" resistance in spite of the existence of "primary groups," while the *Ostheer* fought till the bitter end in spite of their absence, was mainly ideological. The troops of the Wehrmacht had been taught that both on the personal and on the national level surrender to the Red Army was equivalent to giving

oneself up to the devil; individually, because they had been persuaded that the Soviets would kill them outright; collectively, because a Soviet victory would spell the end of civilization, indeed
could only be understood in terms of a universal apocalypse. As
regards the West, Nazi ideology itself was far less extreme, and
could in any case rely to a much lesser extent on the kinds of
popular prejudices against Slavs, Jews (especially the so-called
Ostjuden), and Bolshevism which made the acceptance of its anti-
Soviet propaganda so widespread. Consequently, the individual
soldier believed it safer to surrender to the Western Allies rather
than to the Russians, for while he quite rightly considered his
chances of personally surviving capture in the West much better,
he also did not feel guilty of actively causing his country's extinction by doing so. This was clearly illustrated when during the
last months of the war tens of thousands of German soldiers
strove to extricate themselves from the fighting in the East and
surrender in the West, while those who nevertheless remained
there fought on with suicidal and senseless ferocity. In fact, this
manner of thinking was legitimized even by Nazi propaganda,
which in the last phase of the war increasingly emphasized the
need for the Western Allies to join hands with the Reich against
the "Judeo-Asiatic" hordes threatening to flood the lands of culture and civilization.[12] As Shils and Janowitz themselves conceded, even if they thereby contradicted their own theory,

> the question of the Russians was so emotionally charged, so much
> the source of anxiety, that it is quite likely that the fear of the
> Russians did play a role in strengthening resistance. National So-
> cialist propaganda had long worked on the traditional repugnance
> and fear of the German toward the Russian. The experience of the
> German soldiers in Russia in 1941 and 1942 increased this repug-
> nance by direct perception of the primitive life of the Russian vil-
> lager. But probably more important was the projection onto the
> Russians of the guilt feelings generated by the ruthless brutality of
> the Germans in Russia during the occupation period. The shudder
> of horror which frequently accompanied a German soldier's re-
> marks about Russia was a result of all these factors.[13]

We shall return in a later chapter to some other implication of this
insightful passage. But first let us have a closer look at the actual
destruction of the "primary group" on the Eastern Front. This

process can best be understood first by reviewing the overall tremendous losses sustained by the *Ostheer* and the insoluble manpower crisis into which the Wehrmacht was consequently plunged, and then by examining the impact this situation had on some specific combat units fighting on the Eastern Front.

In preparation for the attack against the Soviet Union the Wehrmacht established fifty-two new divisions, but this considerable expansion was offset by the fact that due to the much larger territories now controlled by the Reich, fully forty-nine formations were employed in occupational duties in 1941, as opposed to only fourteen the previous year. This meant that when the *Ostheer* marched into the Soviet Union with 3,050,000 men organized in 136 divisions (not including its allies), it could leave behind a reserve of only twenty-six divisions, of which fourteen had in fact either not completed their establishment or were still tied up elsewhere. Worse still, the rapid build-up of new formations was achieved at the price of a considerable deterioration of standards, with the result that in June 1941 no less than sixty-six divisions were unfit for combat. These difficulties could have been overcome had the Wehrmacht repeated its former successes. Instead, after just over a month of fighting, the *Ostheer* already lost more men than in the whole Western campaign, and yet was still very far from its primary operational goals. In September the now 142 divisions fighting in the East reported an average decline of close to 50 percent of their initial battle strength, calculated in terms of manpower and equipment, and by November most infantry formations had lost half of their personnel. At this point OKW admitted to having completely exhausted its immediately available manpower reserves.[14]

The Wehrmacht's replacement system was based on a separate *Ersatzheer* (Replacement Army) charged with organizing and training fresh recruits, who were then delivered to the *Feldheer* (Field Army). These conscripts were normally sent to formations recruited from their own regions of origin, and were gradually introduced to their units with the intention of allowing them to fit into existing "primary groups" of veterans, or to form such groups of their own. In June 1941 the *Ersatzheer* numbered 400,000 men, and in order to facilitate the quick delivery of reinforcements to combat formations at the front, 90,000 of these soldiers were organized in so-called "field replacement battalions" which marched directly behind the attacking divisions. But these reserves were quickly consumed, and by August the front needed a further 132,000 replacements. Now the familiar logistical difficulties came into play, for in the fight over transport room manpower reserves were rather low on the list of priorities. Instead, the *Ersatzheer* established another 100 "field replacement battalions" and sent them walking to the front, a rather lengthy procedure considering that the fighting-zone was constantly moving eastward. These were heterogeneous units, for in the rush to reinforce the rapidly dwindling front-line formations, there was no time for consideration of "primary group" ties. By the time they finally arrived at the front, the situation was in any case so bad that these battalions were consumed even more quickly than their predecessors. Meanwhile more radical emergency measures became necessary to expand the Wehrmacht's diminishing manpower reserves. In October five divisions were established from men hitherto exempted from military service due to their importance for industry (*UK-Stellen*); and men from service units in the rear and occupation divisions in the West were earmarked as further *Ostheer* reinforcements. The motley collection of 250,000 new replacements thus created were organized into combat formations around a nucleus of experienced regiments pulled out of the front for this purpose. This measure was once more in contradiction to any former "primary group" practice, for while depriving older combat divisions of a third of their veterans, on the one hand, it established formations of an extremely heterogeneous character, on the other. Nor was this step sufficient to make up for the growing manpower gaps in the East. In mid-December it was thus necessary to send another eight undermanned *Westheer* divisions to the East, accompanied by four so-called *Walküre* divisions, originally intended for use against domestic unrest and unprepared for combat duty. As another desperate measure, the remaining twenty-three divisions in the West were instructed to give up one battalion each, so as to serve as the framework for the creation of a further four to six divisions composed primarily of poorly trained former *UK-Stellen* personnel, thus once more breaking up existing formations and establishing new and highly heterogeneous units.[15]

The manpower crisis of 1941–42 was thus only partially resolved by a series of piecemeal measures which made havoc of the previous replacement system.[16] Consequently the "primary groups" could not be retained to the same extent as in the past, even if the Wehrmacht did its utmost to preserve this tradition. And, since this was only the beginning of a long-term and increasingly deeper crisis, the role of the "primary group" as the basic social unit of combat formations significantly diminished. The inability to overcome the manpower shortage was illustrated by the fact that in July 1942 the number of *Ostheer* troops had diminished by 750,000 men to 2,700,000. Paradoxically, the number of divisions deployed on the Eastern Front actually increased by 43 to 179.[17] This unhealthy development resulted from the attempt to keep formerly homogeneous formations together without being able fully to replace their losses, while using a large proportion of the new reserves to create ever more divisions. In theory, this policy should have at least had the merit of maintaining the old "primary groups," but in practice whereas the "primary groups" were destroyed by casualties, the replacements which did arrive were too heterogeneous to make possible the formation of new "primary groups," and too few to make these veteran divisions once more militarily effective. While the reluctance to disband depleted divisions had to do mainly with Hitler's fear of causing anxiety in the rear, it eventually had the effect of demoralizing the men at the front, for though numerically greatly diminished, these formations retained their old designations and were consequently given tasks far beyond their present abilities. Having seen their old comrades, the members of their "primary groups," killed or wounded, the few survivors of the first winter in Russia could neither form into new groups due to the constant and rapid manpower turnover, nor even briefly enjoy a sense of strength by being sufficiently reinforced. Conversely, the newly established formations, even if they had had any "primary group" ties to start with, came into the war at a stage when these could not be kept for any reasonable amount of time.

While the chronic lack of replacements meant that the manpower crisis could not be overcome and would eventually lead to Germany's defeat, it was the enormous losses at the front which were directly responsible for the physical destruction of the "primary group," as well as for the general deterioration in the professional standard of the troops. During the first six months of the campaign, the *Ostheer* sustained close to 750,000 casualties, rising to over a million, or a third of the entire army in the East, by late March 1942, of whom more than a quarter were either killed or missing.[18] Not only was it impossible to make up for these losses, as OKH reported at the time "all ranks describe the standard of their [replacements'] training and morale [*innere Haltung*] as exceedingly bad." Soldiers with specialized training were particularly scarce. Consequently, in March 1942 only eight divisions, or 5 percent of the entire *Ostheer*, were considered fully prepared for offensive action.[19] Among the officer corps the situation was even worse. Not only were officer casualties proportionately much higher than among the rank and file, but the Wehrmacht had suffered from a lack of officers even before the fighting began. The situation was further complicated by the relatively advanced age of senior and middle-ranking officers due both to the stagnation in promotion in Weimar's Reichswehr, and to the need to re-activate retired officers during the rapid expansion of the officer corps in the 1930s. The harsh living and fighting conditions in the East took a heavy toll on these elderly men. By spring 1942 about half of the divisional commanders in the *Ostheer* had to be replaced. Conversely, efforts to make up for the losses among the junior ranks by instituting crash officer courses produced men who lacked the proper qualifications for filling the positions vacated by their elderly superiors. Yet as circumstances pushed junior officers rapidly along the command structure, platoons and often also companies had to be taken over by NCOs and even privates. The radical transformation in the pattern of officer casualties can be gauged from the fact that whereas during the first two years of the war only 1253 officers died in action, between June 1941 and March 1942 no less than 15,000 officers were killed. Casualties were of course highest among junior combat officers, whose numbers could not be replaced at the same rate. While in July 1941 there were still 12,055 first-lieutenants in the army, by March 1942 their number had diminished to 7276.[20]

Figures concerning the Wehrmacht and *Ostheer* as a whole may suffice to demonstrate that "primary groups" were unlikely to survive the terrible mauling on the Eastern Front. But as we shall see below, among combat units losses were proportionately much higher, and manpower turnover much faster, than these totals would suggest. By observing some of the formations at the front it will become clear that "primary group" must be dismissed as a major factor in sustaining the cohesion of the Wehrmacht throughout the Russian campaign. Let us examine the following examples. The 18th Panzer Division, which crossed the Soviet frontier with 17,174 men and 401 officers, began suffering considerable casualties from the very first days of fighting.[21] Within less than three weeks the division lost some 2300 men and 123 officers, or close to one-third of their original number.[22] Casualties among officers were so high, that by mid-July six battalion commanders were dead or wounded. No wonder that even at this early stage of the campaign, one of the battalions reported that "we have too many casualties. The old spirit is lacking." Toward the end of the month the rifle brigade, which had set out with five fully manned battalions, was left with only 600 men, or a fifth of its initial strength, while the motorcycle battalion lost well over half its men and almost all squad and platoon leaders.[23] By this time the division as a whole had lost 3200 men and 153 officers, or half of those available five weeks earlier. The few officer replacements who now arrived were mainly composed of young and inexperienced men. In order to fill its junior command positions, the division was compelled to commission many of its NCOs, a procedure which made it necessary to give NCO positions to privates. It is clear that within many of the division's combat units there was already a feeling that their "primary groups" had been decimated. One rifle regiment commander lamented the loss of sixty-one experienced officers and over 1000 of his original 2100 men, and went on to say that the meager replacements he had received were not only professionally inferior, but also failed to revive the unit's badly shaken *esprit de corps*.[24]

The independent Grossdeutschland Regiment, a motorized elite unit numbering some 6000 men, found itself in a similar predicament from the very beginning of the campaign. In a typical incident on 5 July, one of its companies was ambushed in a forest and entirely wiped out.[25] This was merely a foretaste of the losses it sustained in fierce forest battles around the Briansk "pocket" in October, where its unit commanders were particularly hard hit.[26] The 18th Panzer, which was fighting next to the GD, was reduced by the end of that month to only 9323 men and 239 officers, even though it had already used up its replacement battalion.[27] Typically, one of its rifle regiments now numbered merely 814 men and twelve officers, which meant that it had diminished to the size of a battalion.[28] As casualties mounted remorselessly in November, there was an increasing feeling that all the "old hands" were disappearing. The divisional pastor wrote at the time in his diary:

> This is no longer the old division. All around us are new faces. When one asks where this or that man is, one is always given the same reply: dead or wounded. Most of the infantry company commanders are new, most of the old ones are gone.[29]

The GD too was rapidly decimated in a series of battles around the town of Tula. In early November one of its battalions was reduced to only eight officers and 359 men, and precisely a month later the remnants of this unit were totally wiped out in a Soviet attack.[30] The 12th Infantry Division, which had marched into the Soviet Union with a combat element consisting of 14,073 men and 336 officers, suffered the same fate. In early November one of its regiments was compelled to disband a whole battalion for lack of manpower, and the division resorted to establishing a convalescent company from troops being treated in the field hospital.[31] By the second week of December the 12th Infantry had lost 4200 men, or close to a third of its initial manpower, and was reduced to 11,351 men and 287 officers.[32]

It was, however, the Red Army's December counter-offensive which finally destroyed whatever "primary groups" may have remained intact. By the end of the month the 18th Panzer reported that the number of soldiers "who actually go into action with a weapon in their hands" had been reduced to a mere 1963 men and forty-three officers, and the whole formation was reconstituted in only four battalions, all of them fighting as infantry.[33] Nor was it the worst hit. In mid-December the 6th and 7th Panzer Divisions, fighting in the same sector as the 18th, were left with only 180 and 200 soldiers respectively. Indeed, the entire LVI Panzer Corps was manned by no more than 1821 "fighters" (*Kämpfer*).[34] The 18th Panzer did receive relatively plentiful replacements in the ensuing winter months, but those were decimated in another bout of heavy fighting as well as from the atrocious living conditions. During the first three months of 1942 alone the division lost a total of 6667 men and 120 officers, a high proportion of whom were victims of disease and frostbite.[35] Yet even excluding the sick, by the end of March total battle casualties since the opening of the campaign reached 9148 men, or over half the division's initial manpower, and 323 officers, or four-fifths of their original number.[36] The 12th Infantry fared little better following its encirclement by the Red Army in February 1942. During the ensuing four months the 96,000 German troops trapped in the Demyansk "pocket" lost a total of 41,212 men, or 43 percent of their strength, while the force sent to break the siege sustained another 12,373 casualties, comprising a quarter of its manpower, within merely four weeks of fighting.[37] This was a heavy price to pay for Hitler's refusal to allow any retreat from a tactically untenable position. As the divisional adjutant of the 12th Infantry reported, the heavy losses made it necessary constantly to re-establish combat units from the battered remnants of destroyed companies and battalions. By sending all available convalescents and service troops into combat, these newly built units somehow held the front. But by now there were no longer any "primary groups" left and, as the adjutant stressed, official unit designations "in no way reflected reality between February and May." Manpower management under such circumstances was so chaotic that by April the division's six remaining battalions were made of such a mishmash of soldiers that one company was said to consist of troops from no less than seventeen different units. Officer casualties were even more distressing. In an attempt to alleviate the situation, the 12th Infantry sent officer cadets to the front, but due to a lethal combination of inexperience and excessive zeal more than half of them were lost within a few days of fighting. As a result seventeen companies had to be taken over by NCOs. The command structure of the division became so disorganized, that according to the adjutant

> due to the merging of units and the need to pull out whole regimental staffs who had remained without any troops, the regiment and battalion commanders were mostly not in a position to exercise any influence upon the placement of officers or to submit any proposals in that regard.[38]

The GD Regiment was hit especially hard, for in the Soviet offensive it was entirely wiped out. Reduced to companies of ten to fifteen men during the retreat from Tula, by the end of 1941 the GD had accumulated a total casualty count of 4070 men, or close to three-quarters of its initial manpower, along with 125 officers, probably more than their original number.[39] By mid-February all that remained of the regiment were two battalions numbering some sixty men each and composed of companies led by NCOs. Soon thereafter the regiment was reorganized into a single rifle battalion, which on 20 February returned from its last operation that winter with no more than three officers and about thirty men and NCOs as the sole survivors of what had been a 6000-man unit.[40] It can thus be concluded that as regards the formations examined here, few "primary groups" could have possibly survived the first winter in Russia, nor could they have played a significant role in sustaining the morale and cohesion of the combat units at the front. Moreover, as can be seen from the soldiers' correspondence, this was the situation in many other combat units as well. Thus, for instance, Harald Henry wrote from the front on 21 December 1941 that he still hoped "to come out even of *this* mess, even if as the last surviving man of the whole company."[41] He was killed the next day. On 24 December Hans Pietzcker wrote: "Dear Mother, it was sad, it was bitterly difficult—of my 36 men only six are still with me," and then hastened to add that "we held out, loyal to our duty and responsibility . . . we were undefeated, we knew our tasks and our orders."[42] Will Thomas wrote on 19 January 1942: "I am the only remaining officer of those who served in the regiment and also the only remaining company commander of those who set out in the autumn." He too was killed the following week.[43]

In the course of the following three years the ever larger losses spelled the repeated destruction of combat units, while the growing overall manpower shortage increasingly hindered the reinforcement of depleted formations and the establishment of new ones. And, as the rate of manpower turnover at the front constantly rose, the creation and preservation of viable and effective "primary group" ties became ever more difficult. On the eve of its 1942 summer offensive the *Ostheer*'s manpower situation remained critical. Even those formations intended to take part in the attack on the southern sector of the Eastern Front had retained only 50 percent of their combat strength, with infantry divisions lacking on average 2400 men, armored divisions between 1000 and 2000. This limited state of preparedness for action could only be achieved by depriving formations stationed along the lion's share the front from any significant reinforcements. Consequently, divisions on the northern sector lacked on average 4800 men each, and in the center as many as 6900, while their combat strengths were reduced to a mere 35 percent of establishment. These depressing figures were of course a direct consequence both of the huge casualties sustained in the first year of the Russian campaign, reaching close to 1,300,000 men (excluding the sick), or 40 percent of the *Ostheer*'s overall manpower of 3,200,000 soldiers, as well as of the general manpower shortage in the Reich which made it impossible fully to replace losses on such a scale. Although the Wehrmacht had called up conscripts earlier than scheduled, and recruited much larger numbers of formerly exempt workers, the lack of men on the Eastern Front could not be reduced to less than 318,000. And, while another 960,000 soldiers were expected to be ready by September 1942, it was stressed that in the event that the summer offensive failed, no further replacements would be forthcoming for quite some time.[44] The Wehrmacht chose to take the risk, and for the second time in a row met with disaster.

Operation *"Blau"* was launched on 28 June 1942, and by mid-September—that is, even before the battle of Stalingrad began taking its toll—over a third of a million men were lost.[45] Following the destruction of Paulus' 6 Army in the cauldron of Stalingrad, the Soviet counter-offensive of winter and autumn 1943, and the abortive German *"Zitadelle"* offensive, losses rocketed to unprecedented levels. Between November 1942 and October 1943 the *Ostheer* sustained well over a million and a half casualties (including the sick), of whom close to 700,000 were permanently lost. As replacements could not keep up with this rate of casualties, no less than 40 divisions were either disbanded or re-grouped into so-called "small divisions," and the establishment figures of the remaining formations were cut almost by half to 10,700 soldiers. Indeed, by December 1943 the *Ostheer*'s overall strength was down by more than a million men to just over 2,000,000 soldiers. In an attempt to make up for this mammoth shortage, the army now greatly intensified the conscription of Soviet POWs and civilians, euphemistically called volunteers, or *Hiwis,* whose number ultimately reached some

320,000 men. While the *Hiwis* were distributed among German formations mainly as replacements for service troops ordered to combat duty, another 150,000 men belonging to Soviet national minorities were organized into semi-independent *Ostlegionen*, though even in this case most command positions were held by Germans.[46] Yet none of these measures, including the transfer to the front in the second half of 1943 of another half a million non-combat troops from the rear, young recruits, women, foreigners, and ethnic Germans, could make up for the increasing losses.[47] In summer 1944 the great Soviet offensive against Army Group Center claimed a monthly average of over 200,000 soldiers and almost 4000 *Hiwis* during the five months between 1 July and 31 December. Indicative of the much greater weight of the Eastern Front even following the Allies' landing in Europe is that in the West the monthly average of German casualties during the same period was just over 8000 men.[48] By November 1944 the *Ostheer*'s total man-power had further declined to 1,840,000 men, and that in spite of the recruitment of yet more *UK-Stellen* personnel and the conscrip-tion of 16-year-old lads.[49] By the end of march 1945 the *Ostheer*'s overall casualties mounted to 6,172,373 men, or double its original manpower on 22 June 1941, a figure which constituted fully four-fifths of the total losses sustained by the *Feldheer* on all fronts since the invasion of the Soviet Union.[50] And yet, among combat units at the front casualties were proportionately much higher still, with a corresponding impact on the formation and life-expectancy of "primary groups."

Throughout spring and early summer 1942 the 18th Panzer Di-vision constantly complained that *"allocation of replacements is in-sufficient,"* that the *"lack of officers* and NCOs is particularly noticeable" and, following the arrival of some reinforcements, em-phasized the "very considerable lack of NCOs and particularly of officers with front-line experience" among them. All that time the division was operating with merely one motorcycle and three in-fantry battalions, supported by a few weak specialized companies.[51] Yet even these remnants of what had not long ago been a well-equipped and fully manned armored division were again destroyed in a fierce Soviet attack in early July. In the first four days of combat alone, no less than 1363 troop and forty-three officer casualties were registered, with one rifle regiment losing all but one of its officers.[52] The effects of such decimation can be gauged from the divisional pastor's diary entries at the time:

> The number of the dead is increasing, the number of the wounded is frightful. In my black book there is already one black cross after another, my whole congregation is almost completely dead or wounded. . . . Then last night we carried the dead out of the trenches and from no man's land. . . . Our cemetery in Bukan has grown tremendously. At first there were only a few graves, now there are already over 400, all in a few days. And how many more will die later in the first aid stations in the rear, or lie in another cemetery? One regiment has brought its sacrifice. The colonel's face has grown terribly thin over these last few days—sleepless nights, turbulent hours. . . . He stands silently in front of the long rows of graves: "There lies my old guard. In fact we too should be there. Then it would all be over."[53]

During the battles of the following winter in the area of Orel, the 18th Panzer was once more severely shattered. Even when it began this round of fighting the division's combat strength stood at less than 3000 men, and by the second week of March 1943 it had only 1994 left in its combat elements out of a total divisional strength of merely 2834 men.[54] The chronic shortage of unit commanders, and their rapid rate of turnover, characteristic by now of all *Ostheer* formations, was well illustrated in a divisional evaluation from early April:

> During the last few days, 8 out of 16 commanders (among whom are all Panzer Grenadier [i.e. infantry] commanders) have been changed; apart from that, the divisional commander and the First Operations Officer are new. . . . None of the commanders have had any battle and command experience in their present position in the *attack*, and only one has had any in the defense. . . . The officer corps at hand is numerically insufficient for either attack or defense. . . . [There is a] shortage of 28 per cent [among NCOs] . . . [who] are good as individual fighters, but most of whom, and particularly the young men, are still not very impressive in their capacity as commanders.

Under such circumstances, it was obviously difficult to form ties of loyalty to the officers, especially considering the parallel constant and rapid decimation of the rank and file. As the above estimation went on to say, the division was now left with four infantry battalions numbering about 290 men each, and its total manpower had been reduced to merely 2440 men, well under a sixth of this former Panzer formation's initial strength.[55]

The 12th Infantry division was hit just as badly. Eleven months after it had entered the Soviet Union, this formation accumulated a total casualty count of 9272 men and 341 officers, that is, two-thirds of the initial manpower and more than the entire divisional officer corps.[56] By this time the division was also some 6000 men and over seventy officers short of its establishment figures.[57] The numerically insufficient and poorly trained replacements tended to arrive piecemeal, which meant that they had to be sent directly into the fighting without the slightest opportunity of either receiving further training or familiarizing themselves with their comrades. As the division reported on 5 June,

> Following the fierce defensive battles during May the gaps torn in the front were filled by inexperienced replacements. . . . The hitherto preferred acclimatization of young replacements in a sector not under attack was not yet possible due to the fact that there was no divisional reserve.

Meanwhile, the number of officers declined relentlessly, and soon the division was faced with a shortage of no less than ninety officers, notwithstanding the fact that its establishment figures had been significantly cut.[58] This situation had a particularly detrimental effect on the 12th Infantry's command structure. By July 1942 the division was left with only five of its nine battalion commanders; consequently, eight company commanders had to take over sectors normally put under charge of more senior officers, while a ninth was detached to another division. This meant in turn that companies were now led by young and inexperienced officers, while platoons were almost invariably led by NCOs, whose own numbers had now greatly diminished because many of those who survived the fighting had already been made into war-officers. As for the rank and file, at this stage only a quarter of them were trained infantry troops; the rest came from non-combat units.[59] This shortage of skilled soldiers, itself the product of heavy losses among combat troops, was also the cause for even greater casualties among men untrained for fighting. As one of the infantry regiments reported, enemy raids against its positions had achieved significant successes due to "the meager numbers of men . . . and the low level of training [which] can only be offset by sending some qualified commanders," who were, however, very hard to come by, as we have already seen.[60] By August the division's losses thus rose to 10,897 men, or close to four-fifths its initial strength, and yet things seemed to be getting progressively worse. Now the divisional commander again pointed out that

> as experience has shown, units which had been replenished by re-placements have sustained particularly high casualties during the first days of combat in a new operation; casualties were that much higher, the larger the infantry units sent to the attack, and the greater the lack of qualified NCOs.[61]

Efforts made by the division to train newly arrived replacements and gradually introduce them to the battlefield[62] failed time and again due to the constant emergencies at the front, which soon made it necessary once more to establish "*Alarmeinheiten*" of untrained service troops whose casualties were probably even higher than those of fresh recruits.[63] With the arrival of winter Soviet attacks inten-sified, and the large-scale decimation of the previous year was re-peated. In early December one of the division's regiments was wiped out within the space of ten days' combat, losing 614 men out of a total of 793, and a few weeks later the powerful "storm battalion" returned from the front with thirty-six men and a single officer, by which time the number of battalions in the division had shrunk from eight to four.[64] These tremendous losses were partially replaced by more fresh replacements, many of whom became casualties just as soon as they arrived from the Reich. The divisional commander repeatedly complained about the lamentable "fighting quality of the young replacements . . . who have received little training with the fighting troops, whether in [offensive] battle or in defensive combat against tank assaults."[65] Yet he had no choice but to make immediate use of them, reporting once more in early December:

> The replacements have gained no self-confidence and fail in the advance. By dodging the tanks they prevent the officers from car-rying out their battle tasks. . . . I caution against a false estimation owing to the large number of replacements, particularly among

companies which have had to be newly established on the front-line.

Evidently shaken by this massacre of untrained young men the commander cabled the corps two days later, emphasizing that

> recruit replacements since early 1942 . . . had not been acclimatized in their new units by training and could not be gradually introduced to battle experience. The good will which they have brought with them has not proved sufficient to overcome the feeling of isolation on the confused battlefield or the terror of the tanks.[66]

But nothing much could be done about this situation. Not only did the troops of the 12th Infantry have little opportunity of forming stable "primary groups," many of them hardly spent more than a few days in its ranks before they were either killed or sent back to the rear in hospital trains. Nor was it possible to form any sort of familiarity with unit commanders. By winter 1942–43 officer replacements were being utilized in a similar manner to the rank and file. Newly arrived and inexperienced officers could no longer be given an opportunity to get to know their units, but were instead sent at their head directly into battle, as unfamiliar with the terrain and the enemy as with their men. This of course led to heavy casualties and was certainly not the best means to strengthen the men's confidence in their commanders.[67]

The Demyansk "pocket" was finally given up by the Germans in the face of an approaching Soviet offensive in February 1943. When the 12th Infantry settled into its new positions along the Lovat river in early March its combat elements numbered only 171 officers and 4822 men, half as many officers and a third as many men as it had mustered at the start of the campaign.[68] The new line did not prove much safer than the old one, and in occasional eruptions of hard fighting units were mauled time and again. In late March, for instance, one battalion returned from a brief but fierce battle with only forty-six men of the original 475, and another was left with sixty-four out of 360.[69] According to the last divisional casualty count available, between 22 June 1941 and mid-October 1943 the 12th Infantry lost a total of 16,112 men, or well over its initial strength, and 527 officers, which constituted 157 percent of their original number.[70] Very little documentation has survived regarding this formation's last days, but it is known that it sustained heavy losses in the defensive battles around Vitebsk in winter 1943–44,[71] and was then completely destroyed in the Soviet summer offensive of June 1944.[72] This could mean that during its three years on the Eastern Front the 12th Infantry lost as many as twice its initial number of men and about three times as many officers. This scale of casualties, and the consequently rapid rate of unit destruction and manpower turnover, thus clearly hindered the formation of "primary groups" and greatly limited the impact of personal ties between unit members on motivation and cohesion in battle.

Unlike the formations surveyed above, the newly established Grossdeutschland Division survived till the very last days of the war thanks to a constant flow of replacements. It thus serves as an excellent example of a highly motivated division whose astonishingly high casualties precluded the preservation of anything but the most limited degree of "primary group" ties. The GD entered the German summer offensive of 1942 with some 18,000 men and 300 officers, and even in the first month of fighting lost over 2000 men and 114 officers, or close to 40 percent of their original number.[73] But it was only after the division was transferred to defensive positions on the central sector of the front that casualties reached truly alarming proportions. In September alone one of the infantry regiments lost close to 1400 men, with one battalion being entirely wiped out and many companies reduced from 140 to less than fifty men. Officer losses were especially severe, including a regimental commander and two battalion commanders with most of their staffs killed by direct artillery hits on field headquarters, an event which was to recur frequently over the next three years.[74] In November things became even worse. One battalion was destroyed within twenty minutes while charging into a Soviet artillery barrage, losing virtually all its officers from platoon leaders to the battalion commander. The fate of a neighboring unit was described by one of its members:

> Completely isolated, [the second platoon] fought . . . to the last man. . . . As both our company commander, Lieutenant Rupp, and his Sergeant-Major Sonnenberg were missing after these battles, and the soldiers of the first and second platoons had been mostly wounded or killed, our old "sixth" [company] was completely annihilated but for five men. During the next few days . . . the battalion fought as a small battle-group.

Three days after this diary entry was made, the battalion in question was reduced to the size of a fully manned platoon.[75] These losses compelled the GD to establish *"Alarmeinheiten,"* with the familiar result of even higher losses among these untrained service troops.

By the end of the month at least another two battalion commanders were killed, and in December a whole infantry regiment was encircled by Soviet tanks and totally wiped out, an event described by the divisional operations officer in a letter to his wife:

> The most frightful incident occurred when a regiment commander bid farewell from me on the radio. He described how they had been surrounded, fighting hard, and yet the Russian came ever closer, a withdrawal was of course forbidden. "Here we now remain," he said, "my men are fighting like lions, but soon it will all be over, we are ready, please stay a little longer on the line, I would still like to speak to you—now the tanks are coming"—then the connection was cut. The colonel is dead, his staff have fallen, only one officer and one man have succeeded in breaking out.[76]

As the fighting went on, yet another battalion was reduced to four officers and thirty-five men. The manner in which one unit after another was destroyed in futile counter-attacks can be gauged from the description of one such operation by an infantry company:

> The heights were taken in spite of heavy defensive fire. But in the battle for the top bunker one man after another was killed. Lieutenant Weiss and his five remaining men could no longer take the last position. And so they left the heights held by the Russian with the sole achievement of having had the whole company smashed to bits.[77]

The company to which the trooper who wrote this passage belonged was itself decimated, having lost six commanders since July and numbering by the end of December just one officer, two NCOs, and eight men. Nor was this exceptional; a neighboring company had been reduced to one officer, one NCO, and 16 men.[78] Indeed, within the preceding four months the remaining infantry regiment had lost 2674 men, which constituted two-thirds of its manpower.[79] And, as replacements could not keep up with these losses, most companies had only one officer left, while platoons were led by NCOs and even privates.[80] As the division went on fighting almost without pause until spring 1943, by April its total casualty count reached 10,365 men, just under two-thirds of its original strength merely ten months earlier, while 375 officers, or 125 percent of their number in June 1942, were either dead or wounded.[81] That such destruction of combat units was not exceptional at the time can be seen from numerous soldiers' letters. Thus, to cite only one example, on 15 September 1942 Martin Linder wrote from the Eastern Front:

> Since 28 June 1942 alone our company has sustained 190 casualties in wounded and killed. As wounded are considered those who must be sent to hospital—many of the lightly wounded remain either directly with the combat troops or recover among the service units. Of the 190 [casualties] 34 were killed and one can calculate that a further 10 to 15 have either died in hospital or are certain to die of their wounds in the future.[82]

Returning from a brief leave at home, Linder wrote on 28 November 1942 that while he had been away "[t]wo-thirds of my platoon have become casualties." [83]

In the few months preceding the last large-scale German offensive in the East, operation *"Zitadelle,"* frantic attempts were made to reinforce the *Ostheer*'s formations. But both the inadequacies of the preparations, and the paroxysm of destruction which seized the front once the attack was launched, forcefully demonstrated that the terrible manpower drain did not merely prevent any meaningful resurrection of the Wehrmacht's "primary groups," but was rapidly bleeding the Reich white to the extent that it would soon have no more men fit for army service left at all. In fact, as of summer 1943 fighting units sustained such a high rate of losses, that they were eventually wiped out one after the other, though fighting until the very last moment with dwindling replacements gathered from every conceivable source. Here the 18th Panzer serves as a particularly good example. In spring 1943 this formation's combat elements numbered 124 officers and 3782 men, less than half its original strength. Yet even these figures were reached by ordering all service troops belonging to the classes of 1906 and 1907 to combat duty, which meant that a large proportion of the division's fighting units were now made up of ill-trained men aged over 36 years.[84] In order to bring the division to strength in view of the approaching offensive, in May 1943 it finally received eighty-nine officer, 130 NCO, and 2571 troop replacements.[85] But once again this uncommonly generous reinforcement largely reflected the deterioration in the Wehrmacht's overall situation. It soon turned out that half of the new officers were in fact merely cadets, and the other half were elderly, inexperienced, and poorly trained. Similarly, the NCOs, quite apart from being numerically insufficient, were also professionally disappointing.[86] Nevertheless, as the division entered the battle of Kursk with four infantry battalions and three support battalions, it was relatively far more powerful than most other German divisions at the time.[87] In the course of the next four months, however, the 18 Panzer sustained such heavy losses that it was finally compelled to disband altogether. On 11 July, less than a week after *"Zitadelle"* was launched, the 18th Panzer's combat elements still numbered 5266 men and 157 officers; twelve days later only 890 men and less than thirty officers were left.[88] This tremendous decimation prompted the divisional commander to order all available service troops to the front, not least because he knew that "the higher leadership will set the division tasks which cannot be carried out with the combat strengths reported hitherto." Yet considering previous screenings of service units, it is clear that these men were entirely unfit for combat duty.[89] A week later one of the division's two rifle regiments was reduced to merely 127 soldiers, and by 27 July it had only one officer company commander left.[90] All in all the division lost 3198 men and 145 officers in July alone, about half its men and almost all its officers in a single month of fighting. At this stage both rifle regiments put together had 482 men left, and all but four companies were led by NCOs. Whatever "primary groups" had formed in the months leading to *"Zitadelle"* were thus wiped out within a few weeks of fighting.[91]

In early August the division was reinforced and its combat strength raised to 113 officers and 3643 men. But the replacements were described as invariably elderly and poorly trained men. By the end of the month the 18th Panzer reported a further toll of 1249 casualties, to which were added another 825 wounded and 685 missing soldiers lost during the preceding two months and not previously reported due to the chaotic conditions at the front. The only way the division could continue operating was by attaching to its units men from disbanded or routed formations roaming its area of operations. Officers were now created by rapidly promoting NCOs, whose own dwindling numbers were similarly raised by promoting privates.[92] Yet as the fighting went on, by mid September the division had only two infantry battalions left, numbering together 308 men and supported by 91 pioneers and 118 replacements, and by the end of the month the division counted another 1181 casualties, or half of its manpower four weeks earlier.[93] The intensity of the fighting can be gauged from noting that between 4 July and 30 September a total of 218 officers were lost, well over their number at the beginning of *"Zitadelle."* These officers included 10 battalion, 83 company, and 85 (officer) platoon commanders. All in all, since 22 June 1941 the 18th Panzer sustained 695 officer casualties, or 173 percent of their original number, a third of whom were lost in the last three months of its existence. By the time the division was finally disbanded on 10 October 1943, it counted 17,001 rank and file casualties, which was just as many men as it had had at the start of "Barbarossa."[94] One can conclude that in spite of fighting fierce battles until the end, the 18th Panzer's troops could not have maintained more than the most ephemeral "primary group" ties, nor felt much personal loyalty to their unit commanders, during the major part of their service in the Soviet Union.

The GD Division's survival until April 1945 allows us to examine the relationship between the extremely high casualties of those last two years and the no less remarkable determination of German troops in the face of certain defeat.[95] For this formation too, the battle of Kursk meant terrible losses within the space of a few weeks; but it also spelled the acceleration of the cycle of destruction and reconstruction which continued well into the last months of the war. Even during the first days of *"Zitadelle"* the commanders and staff officers of both infantry regiments, and of another three battalions, were almost all either killed or wounded.[96] By mid-August the average strength of the infantry companies had dwindled to some twenty-five men, many of them now led by NCOs for lack of officers. Here too the practice of making immediate combat use of replacements proved very costly; in one such instance a group of young recruits who had just arrived at one of the companies were sent directly to the battlefield and within a few hours six were dead and four wounded.[97] By early September most companies were hardly larger than squads. The sixth Grenadier company was composed of five men led by a lance-corporal, the seventh of five led by a corporal, the entire third Grenadier battalion had three officers and 29 men, and the second battalion had three officers and 22 men.[98] As the GD retreated to the west bank of the Dniepr in late September, it reported a total casualty count of 7347 men within the preceding six months, bringing its overall losses since June 1942 to 17,712 men and 590 officers, or almost as many men as it had set out with and double the number of officers.[99] But the fighting went on without pause. By November the pioneer battalion lost all but one of its officers, while the two infantry regiments put together had only 250 men left, with one battalion reduced to twenty-seven soldiers.[100] Even after some replacements arrived in early 1944, it was reported that battalions "still had 100 men, a considerable force at the time. There were two companies of 40 men each, and a reserve of about 20 men."[101]

Some insight into the rapid turnover of combat unit commanders can be gained from the following example. On 26 July 1943, Second-Lieutenant Heinz, commander of the sixth Grenadier company, was killed and replaced by an NCO; on 7 August Second-Lieutenant Ahlfeld took charge of the company, but was killed four days later. Second-Lieutenant Pfau took over on 13 August and was wounded after eight days. His replacement, Second-Lieutenant Fritz, was wounded on 2 September and an NCO had to command the unit. On 17 September Lieutenant Wiebe took over, but on 6 October Pfau returned, having recovered from his injury, only to be wounded once more two weeks later. Second-Lieutenant Hegemann, who took over on 3 November, was reported missing after three days, and his replacement, Second-Lieutenant Michaelis, was wounded in turn ten days later, relinquishing company command to Lieutenant Saalfrank. This was an average rate of a new company commander almost every week. Battalion commanders did not fare much better. On 14 October Captain Knebel, commander of the second Grenadier battalion (to which the sixth company belonged), was replaced by Captain Kraussold, who within a couple of weeks was replaced himself by Lieutenant Konopka. On 17 November Konopka was wounded, but Captain Kraussold, who again took over battalion command, was seriously wounded only hours later. Now Captain Weizenbeck assumed command, but less than two weeks later he too was wounded. His replacement, Captain Krambeck, was also wounded a few days later and had to relinquish command to Captain Mickley.[102] Under such circumstances it would indeed be hard to believe that soldiers could develop much personal familiarity with and loyalty to their ever changing commanders, quite apart from their own rapid manpower turnover. In a letter sent on 12 July 1942 from Sevastopol, Friedrich Reinhold Haag, a company commander in another formation, described the effects of such rapid manpower turnover:

> I have experienced again how difficult it is to lead a company into a battle and to sacrifice men while hardly knowing any of them. Then they fall right next to you and one of them cries perhaps: "Herr Leutnant, be sure to write home"—and you don't even know what his name is.[103]

Whatever remained of the GD following these winter battles was decimated during the long retreat to Rumania in the wake of the Red Army's great offensive in the Ukraine in March 1944. It was in this chaotic withdrawal that the sixth Grenadier company, not unlike many other combat units, lost three commanders, of whom one was an NCO, within the space of only three days.[104] Having reached the Rumanian frontier, however, the division was rapidly reinforced, and in early April boasted of four infantry battalions numbering some 400 men each, along with other support units. But once more there was no time for any ties to develop between its new members and the few surviving veterans, for following another Soviet attack by 10 May the battalions were reduced to an average of 65 men, having lost close to four-fifths of their troops within less than a month.[105] Another batch of replacements was similarly mauled in defensive warfare in June, and when for the third time in three months the GD re-established its units, the fresh replacements were mainly composed of fifteen- and sixteen-year old lads taken directly from the *Hitlerjugend*, along with some elderly Great War veterans.[106] This extraordinary rate of destruction and replacement continued after the division was transferred to East Prussia later that month. In heavy fighting during the first half of August

1944 companies were again reduced to an average of twenty men.[107] Officers were as always particularly hard hit: the four companies of the first Grenadier battalion registered twelve changes of command in the space of merely nine days. Indeed, by the time this Battalion retreated to Memel in the first part of October, it was composed of eighteen men and led by a sergeant, himself wounded a few days thereafter.[108]

Now for the last time the GD was re-established. Evacuated by sea from besieged Memel, in November 1944 the remnants of the division arrived at Rastenburg in East Prussia, where substantial reinforcements brought up their strength to about 10,000 men.[109] The majority of these soldiers had had no previous experience in the GD, nor did they have much of an opportunity to gain any now, for when the division went back into action in early 1945, it was rapidly torn to shreds in the final mammoth Soviet offensive. Soon cut off from the main bulk of the *Ostheer*'s retreating formations, between 15 January and 22 April the GD sustained the astonishing total of 16,988 casualties, constituting 170 percent of its manpower at the beginning of 1945. By now the 4000 GD survivors were fighting desperately in the Samland peninsula, filling their ranks with the remnants of other German formations scattered in the area.[110] Even at this last stage of the war plans were drawn to extricate the division from East Prussia and re-establish it somewhere in the ever contracting Reich, but time finally ran out.[111] In mid-April the Red Army attacked once more, and following a few days of fierce fighting, the last 800 GD soldiers swam across the bay to the *Frische Nehrung* land-strip where they were picked up by a ship which took them to Denmark. Most of these men, who ended up in British captivity soon thereafter, had been in the GD only a few weeks or months.[112] Since it first went into action in June 1942 until the end of the war, the GD lost about 50,000 men and 1500 officers, or close to three times the number of men and five times the number of officers it had had upon establishment.

From everything we have seen above, one cannot avoid the conclusion that throughout the war in Russia the "primary groups" of the *Ostheer*'s combat units could not have survived more than a few weeks at a time under battle conditions, and could thus not have played a significant role in the cohesion and motivation of the main bulk of the Wehrmacht's land forces. Similarly, personal loyalty to officers could also not have been a significant motivating factor due to their immense casualties and the resulting extremely rapid turn-over of unit commanders. The question that should now be asked is what did make for the German army's remarkable cohesion? In the next chapter we shall suggest that military discipline greatly contributed to keeping the men at the front, but that this was achieved only by profoundly perverting the nature and meaning of discipline, a process which had an important impact on the troops' conduct and state of mind, ultimately leading to a widespread bru-talization of combat units.

3

The Perversion of Discipline

Harsh military discipline had a long tradition in Germany, stretching back to the days of corporeal punishment in the old Prussian army.[1] Just like organization and planning, discipline and obedience were the trademarks of the German military long before Hitler came to power. Frederick the Great's line tactics and the oblique attack, the elder Moltke's massive concentration of forces by making precise and innovative use of trains, the clockwork-like realization of Schlieffen's huge "swinging door" strategy—all seemed to prove the efficacy of these German virtues on the battlefield.[2] It was consequently seen as only natural that the Wehrmacht too had maintained the rigid disciplinary system of its predecessors. Indeed, while its clever combination of the old military tradition with modern fighting techniques can be regarded as one of the keys to the Wehrmacht's astounding military successes, the strict obedience demanded from the troops, and the draconian punishments meted to offenders, doubtlessly played a major role in maintaining unit cohesion under the most adverse combat conditions. Nevertheless, it must be emphasized that in the Third Reich, and especially during the war, the theory and practice of martial law in the Wehrmacht underwent fundamental and crucial changes, which both reflected and enhanced the overall transformation of the army's character, and were responsible not only for the troops' steadfastness on the battlefield, but also for their profound brutalization.

The widespread, though temporary breakdown of discipline in the Kaiserheer during the final stages of the Great War left a lasting impression on the men who were to serve as the architects of the newly established Wehrmacht. Just as Hitler feared that the demands of another total war might destabilize his regime and bring it down as rapidly as the Kaiser's, so too the generals feared a repetition of the humiliating scenes of soldiers' councils refusing to obey their officers.[3] The new Wehrmacht was supposed to do away with the social barriers between officers and men, on the one hand, and to demand "blind" obedience and unquestioning loyalty from the troops, on the other. Soon enough the same "blind" obedience, this time directly to the Führer, was also demanded from the generals and staff officers, who had long nurtured a tradition of independent and critical thinking, quite contrary to the stereotypical view of the German military. Ideologically founded on the myth of the so-called *Frontgemeinschaft* of 1914–18, and on the related belief that only total spiritual commitment would enable one to withstand, if not actually celebrate, the horrors of modern war, the new Wehrmacht shared numerous common roots with National Socialism, much of whose own vocabulary and imagery had been lifted directly from the realities and fictions of the Great War.[4] Thus from the very beginning military discipline in the Third Reich's army was closely tied to the ideological determinants of the regime, not merely by coercion but very much by choice and affinity. Traditional norms of discipline and obedience were consciously and willingly adapted to fit the political concepts deemed necessary for improving the effectiveness of the military machine and the cohesion and morale of its troops.

The politicization of discipline went hand in hand with a politicization of the army as a whole.[5] Attempts to keep some of the old traditions whilst permitting the ideological penetration of the military were bound to fail. The army's top priority was combat performance, one of whose preconditions was assumed to be a new and more demanding disciplinary system based on and legitimized by the introduction into martial law of Nazi legal and moral concepts and norms of behavior. This voluntary legal *Gleichschaltung* inevitably led to a profound perversion of discipline in the Wehrmacht.[6] To be sure, draconian punishment did instill into the troops fear of their commanders and induced them to carry out their orders even under extremely difficult conditions; it also greatly brutalized the men. Fearful of their commanders, and unable to defeat the enemy, the troops turned against the occupied civilians and prisoners. This was a gradual process, dependent both on the ideological stance of the regime toward each of its enemies, and on the degree of resistance met by the Wehrmacht. As for the soldiers, prejudice, fear, and brutality became closely entwined. During the Wehrmacht's rapid successes in the first two years of the war traditional discipline normally sufficed. While the troops were not punished with inordinate cruelty, the enemy's civilians and prisoners were also treated decently, and soldiers stepping out of line in their conduct toward the occupied were severely punished. However, as was particularly the case in Poland, this did not apply to political and "racial" categories deemed by the regime, and thus also by the Wehrmacht, as undeserving of the accepted rules of war. The potential of ideologically determined brutality revealed in autumn 1939 became the rule with the invasion of the Soviet Union. Here the perversion of discipline could be seen on three, closely tied levels: within the ranks of the army, breaches of combat discipline were punished with unprecedented harshness and contempt for life; conversely, soldiers were ordered to commit "official" and "organized" acts of murder and destruction against enemy civilians, POWs, and property; and, as a consequence of the legalization of criminality, the troops soon resorted to "wild" requisitions and indiscriminate shootings explicitly forbidden by their commanders. In stark contradiction to the harsh combat discipline, however, the troops were rarely punished for unauthorized crimes against the enemy, both because of their commanders' underlying sympathy with such actions, and because they constituted a convenient safety valve for venting the men's anger and frustration caused by the rigid discipline demanded from the men and by the increasingly heavy cost and hopelessness of the war. Thus a vicious circle was created whereby the perversion of discipline bred increasing barbarism, which in turn further brutalized discipline. That the army's murderous policies in Russia only swelled the ranks of the partisans, justifying even greater annihilation and destruction, merely reflected the penetration of Nazi legal perversity into the Wehrmacht, not unexpectedly accompanied by the irrational and nihilistic modes of behavior typical of the regime as a whole.[7] From the army's point of view, quite apart from implementing Hitler's policies, the only binding laws in its anti-Bolshevik "crusade" were those which ensured the preservation of its own cohesion and effectiveness.[8] Whatever served this goal had to be ruthlessly enforced, and whatever stood in its way had to be just as ruthlessly suppressed. Yet while fear of one's commanders normally sufficed to keep even the most faint-hearted in their positions, when terror of the enemy became even greater, breakdowns did occur. The growing incidence of local disintegration demonstrated that discipline was not enough; but the fact that such occurrences could be isolated and did not substantially influence the army's cohesion as a whole showed that there were even deeper causes for the troops' willingness to go on fighting.

The aim of military discipline is of course to ensure that soldiers follow a certain set of rules and regulations which provide for an army's cohesion and combat effectiveness; it is also a reflection of that army's self-image. In this sense both the content and the means of enforcement of such rules and regulations indicate the underlying ideological concepts of the army. At the same time, the aggravation of brutality by the troops was also a reflection of the social influence of Nazism and, more specifically, of the declared aims of the war, especially against the "subhumans" in the East. The transformation in the Wehrmacht's self-perception can clearly be observed through the judicial and practical changes in its disciplinary system. This process can be illustrated by a few examples. On 25 October 1939, less than two months after the invasion of Poland, the commander in chief of the army complained in a circular to all units: "The achievements and successes of the Polish campaign cannot allow one to overlook the fact that a part of our officers lack a firm inner bearing [*feste innere Haltung*]." Von Brauchitsch was particularly worried about cases of illegal exactions and requisitions, embezzlement and theft, maltreatment of subordinates, drunkenness, insubordination, and rape. He stressed that the conduct of officers would have an immediate effect on the soldiers, and in a special supplement outlined those areas which were of especial significance for the officer's "inner bearing" and appearance. The commander, he maintained, should set an example to his men by his readiness to fulfill his duty and by his sense of responsibility, but he should also care for his subordinates in their daily affairs; he should enforce strict but just discipline; he may never be drunk, and must take care that his "exterior bearing," his uniform, remain in faultless order, even in the most difficult battle situations. Moreover, the officer's wife too must not only be a good spouse to him and a devoted mother to his children, but should also assist him in his professional capacity by serving as an example to the rank and file.[9]

Orders of this sort, stressing the officer's bearing, appearance, and care for his men, had a distinctly traditional air and wording. Formation commanders similarly tried to instruct their officers in the old spirit of chivalry and paternalism. In November 1939 the commander of the 12th Infantry Division warned his officers not to enter public bars or public dance halls in uniform, not to be seen with "ladies" who were not absolutely beyond reproach, and to take more care with uniforms and salutes.[10] Finally, in December, von Brauchitsch clarified to his officer corps the close link between "outer and inner bearing." Blending the traditional asceticism of the officer corps with the new anti-materialist terminology of the Nazis, the commander in chief of the army wrote:

> It is this historical legacy whose preservation should be the highest duty of anyone who has had the honor of wearing the officer's coat. Bravery in the face of the enemy alone is not enough. It is just as important to avoid all temptations in daily life. Notwithstanding all idealism, material things force themselves greatly upon one's thoughts; the officer must therefore serve as an example of abstention precisely in this sphere.

There were indeed reasons to worry about the officers' moral outlook, as the supplement to Brauchitsch's order clearly revealed. Here was a long list of officers tried and severely punished for offenses such as theft, drunkenness, misconduct and reprehensible sexual behavior.[11] And, as could be expected, the troops' discipline during the Polish campaign and the subsequent occupation of that country was likewise far from satisfactory. As early as 29 September 1939 the 12th Infantry Division appealed to its troops to take more care of their appearance and soldierly conduct,[12] and two days later Army Group South noted that "discipline . . . among all *rear units,* as well as among many newly established units . . . has reached such a low point, that it must be dealt with by employing all possible measures."[13] The steep rise in disciplinary infringements during the occupation of Poland was evident from the court-martial records of the 12th Infantry. Whereas in September this formation tried only seventeen men, in October the number almost doubled to thirty-two, reaching a peak of sixty-three trials in November. By then the division was back in the Reich, but the after-effects of the occupation were so prolonged that even in December 1939 and January 1940 an average monthly rate of fifty trials was maintained.[14] Growing nervousness among senior officers regarding the detrimental effects this disciplinary deterioration might have on battle performance and cohesion led to increasingly harsh punishment. On 8 November 1939, for instance, 4 Army announced that it would severely punish men found guilty of such offenses as absence without leave, desertion, plunder, and disobedience, and went on to stress that in particularly serious cases it would not hesitate to make use of the death penalty.[15]

The occupation of a foreign country is always a threat to military discipline. But in the case of the Wehrmacht the situation was further complicated by the underlying ideological determinants of its own disciplinary system, as well as by the criminal activities of the SS in areas directly under the army's control which, however, not only did it lack authority to stop, but was in fact expected to support both practically and morally. Brauchitsch himself, in spite of his respect for the officer corps' traditional values, was simultaneously at pains to persuade his officers and men of the need to act and feel as the true representatives of Hitler's new Reich. As early as 19 September 1939, the commander in chief of the army warned in a so-called "Leaflet for the Conduct of German Soldiers in the Occupied Territory of Poland" that the troops should expect to be confronted with "inner enmity" from all civilians who are not "members of the German race." Moreover, he went on to say, "The behavior toward Jews needs no special mention for the soldiers of the National Socialist Reich."[16] In Poland the status of the occupied had nothing to do with international law and conventional military practice, but was determined strictly according to biological and political criteria. So as to dispel any feeling that such instructions might conflict with the army's traditional values, on 25 October, in the very same order that called upon officers to take care of their "outer and inner bearing," Brauchitsch explicitly forbade "all criticism of the measures of the state leadership" being carried out at the time by the SS in the occupation area, and called for "strict silence" and "avoidance of all gossip and the spreading of rumours." Indeed, even the officer's wife was expected to set an example of ideological conformity. An incident which involved two senior officers' wives, one of whom had made "derogatory remarks about the political leadership and other defeatist comments of all sorts," and the other who had denounced her to the Gestapo, furnished the commander in chief of the army with the opportunity to air his views on the subject. "I have already pointed out numerous times," he wrote in an order circulated to all units,

> that the officer corps and thereby the officers' wives must set an example of faith [*Bekenntnis*] in National Socialism and in constantly behaving according to the dictates of that conception [*Anschauung*]. Any offense against this is a severe injury to the appearance of the officer corps.[17]

Discipline was thus increasingly becoming a political issue. To be sure, long before the outbreak of war expressions of political criticism were a punishable offense in the army; but the troops were by and large kept away from the practical implications of Nazi policies. In Poland things took a radical turn, and although the Wehrmacht was not officially and directly involved in the murders committed by the SS, they often occurred in sight of the troops, indeed under their protection. The traditional disciplinary code and value system did not allow for such crimes, and the officer corps might have been expected to be morally revolted by them, if not indeed physically to prevent their continued perpetration. Yet from the regime's point of view the practical implementation of its ideology was nothing but the raison d'être of the war. The army could thus either rebel against the regime or adapt itself to a new set of norms and values. Significantly, choosing the latter alternative also meant that the troops' conduct would fundamentally change.[18] This was clearly understood by General Blaskowitz, the commander of the occupation forces in Poland, as illustrated in his memorandum of 6 February 1940. Warning against the beginnings of partisan activities against the Wehrmacht, Blaskowitz went on to say:

> It is incorrect to slaughter a few ten-thousand Jews and Poles, as is happening at the moment; for this will neither destroy the idea of a Polish state in the eyes of the mass of the population, nor do away with the Jews. On the contrary, the manner in which the

slaughter is carried out causes great harm, complicates the problems
and makes them much more dangerous than they would have been
if premeditated and purposeful action were taken.

This was an ambiguous statement, for it could also be read as a
recommendation to kill more people in a more orderly and disciplined manner, rather than to refrain from the butchery altogether.
Indeed, this was the reason that such pogroms as *"Kristallnacht,"*
or even the murder campaign of the *Einsatzgruppen,* were rejected
in favor of the organized, systematic, disciplined killing in the gas
chambers. Concerned more with the military consequences of the
massacres, however, Blaskowitz maintained that they would merely
assist the enemy's propaganda, while forcing the Poles to join forces
with the Jews against the Germans. Furthermore, he stressed:

> There is hardly any need to mention the effect on the Wehrmacht,
> which is compelled passively to observe these crimes, and whose
> standing particularly among the Polish population will be irreme-
> diably harmed.

Most perceptively, Blaskowitz realized already in this early stage of
the war that even the relatively passive role played by the Wehrmacht
in these crimes was bound to have the most serious long-term effects
not only on the soldiers, but on German society as a whole:

> However, the worst damage which will be caused to the body of
> the German nation by the present conditions is the boundless bru-
> talization and the moral depravity which in the shortest time will
> spread like an epidemic among the best German human material.

This, Blaskowitz understood, would be the consequence of "legalizing" criminality and thereby making it virtually illegal to act
against what by every accepted standard of human decency and
morality was clearly criminal:

> When high officials of the SS and the police call for atrocities and
> brutalities and publicly praise them, then within the shortest spell
> of time only the brutal will rule. With astonishing speed men of
> the same sick leanings and character will come together, in order
> to give full vent to their beastly and pathological instincts, as is the
> case the Poland. There is hardly any way to hold them in rein; for
> they must rightly feel themselves officially authorized and entitled
> to any atrocity.

The only way to protect oneself from this epidemic is by bringing the guilty and their followers at the greatest speed under military command and jurisdiction.[19]

In fact, however, rather than bringing the SS under military jurisdiction, martial law increasingly remolded itself to fit the Nazi concept of "racial" and political justice. As Blaskowitz had predicted, the occupation of Poland was only the beginning of a process of brutalization which soon spread throughout all ranks of the army. Yet this was not an entirely irreversible development, as the radically different course of events in the West proved less than a year later. Here the SS was not allowed the same freedom as in Poland, due both to political considerations and because the nations of Western Europe ranked much higher on the Nazi "racial" scale than the Slavs in the East.[20] To be sure, the brutalization of the troops in Poland did have an effect on their conduct in the West; but not only were they less plagued with prejudices about the inhabitants of these more "civilized" countries, this time their officers stepped in with harsh, occasionally even brutal disciplinary measures.[21]

Before turning to the Eastern Front, the main theatre of the war, it is thus instructive briefly to examine the manner in which commanders of combat formations in the West reacted to acts of looting and violence by their troops against the civilian population. The fighting had been in progress for only two weeks, when the commander of the 12th Infantry Division warned his troops to avoid unnecessary plunder, though significantly he also authorized unit commanders to take from the inhabitants whatever they deemed necessary:

> "Living off the land" must be under the unit commander's supervision. The dividing line between plunder and procuring the essential needs must be strictly observed. The troops should not live badly, but must also neither dissipate nor enjoy their provisions too greatly, particularly as regards alcohol.[22]

Even at this early stage of the campaign there was clearly a great deal of looting and drinking going on, and senior commanders became increasingly determined to tighten their control over the troops' conduct. Yet it must be stressed that it was the official authorization given to combat units to "live off the land" which so greatly enhanced the potential of outright plunder; only the forceful intervention of high-ranking officers, backed by the army's rules and regulations, prevented this campaign from degenerating into one of robbery and murder. The Wehrmacht's troops showed themselves capable of a high degree of brutality even when under strict orders to observe the laws of war; but in the West commanders could still be shocked and distressed by such crimes, as was obvious from an order issued by the commander of 4 Army, von Kluge, two days after the end of hostilities:

> Trial proceedings show a frightening increase in rape offenses. The detailed circumstances are often downright heinous. The discipline and appearance of the Wehrmacht in the occupied territory will face grave dangers in view of these crimes.[23]

The corrupting effects of occupation, the example given by the SS in Poland, and the highly ideological context in which the war was being fought, all made it extremely difficult to control the troops. Nor were the officers always blameless. In early October 1940 the 12th Infantry noted that the German police had found plundered Polish and French goods in soldiers' homes; these goods, it was said, had been "distributed in the *presence of senior officers* to officers, NCOs, and soldiers."[24]

In order to prevent the spread of such conduct the army resorted to very harsh punishment. By late October 1940 the 12th Infantry sentenced to death five soldiers found guilty of rape and armed robbery, and sentenced another forty-two to over one year's imprisonment.[25] But as the occupation went on, discipline continued to deteriorate. Whereas during the first three-quarters of 1940 a monthly average of only eighteen men were court-martialled, in the last quarter of the year the average leaped to over thirty, reaching as many as forty-five trials per month in the first quarter of 1941.[26] The divisional court martial blamed the long "rest period" on this disturbing increase in breaches of discipline, and the divisional commander similarly argued that "the long rest period in a rich land" had led astray "many soldiers whose character in not strong enough to resist temptation."[27] Yet judging by the nature of the offenses, it seems they were just as much the product of the brutalizing effects of war, occupation, and an ideology which allowed one severely to maltreat the enemy and yet retain one's inherent moral superiority. Apart from the more common breaches of discipline such as absence without leave, theft, and insubordination, the division noted a rise in cases of plunder, drunkenness, brawls with officers and civilians, as well as sexual attacks on women, children, and even intercourse with animals.[28] It should be noted, however, that troubling as this situation was, it remained both quantitatively and legally within the bounds of what could be expected under conditions of military occupation. While the number of trials stabilized at the relatively low monthly average of thirty, on the one hand, it seems that by and large soldiers who committed offenses against members of the civilian population were brought to justice and severely punished, on the other.[29] There were only two, though highly significant exceptions: those perceived as the Reich's political enemies, be they former German citizens, foreign political opponents, or resistance fighters, and those labelled as the German *Volk*'s biological enemies, and especially Jews, were treated not only by the SS, but also by the Wehrmacht, in an entirely different manner and could not expect any legal protection. Here military discipline showed its capacity not only to prevent crimes, but also to legalize them. As early as 21 June 1940 the 12th Infantry was issued the following order:

> Prisoners who are Germans belonging to the Reich [*Reichsdeutsche*] (including areas annexed to the Reich) and Czech citizens, since they also count as members of the German Reich, as long as they are so-called emigrants, are to be shot after their identity has been established. The execution should take place in POW camps.[30]

Men belonging to this category were of course either political refugees or Jews who had escaped Nazi Germany and had been denied citizenship by their host countries. But while in the West this was at least initially a limited phenomenon, which moreover had little impact on the local population at large, in the East it became the basis for the Wehrmacht's occupation policies.

The German army invaded the Soviet Union equipped with a set of orders which clearly defined "Barbarossa" as a war essentially different from any previous campaign, a "war of ideologies" in which there were to be "no comrades in arms." It is the fundamental contradiction in terms encapsulated in what have come to be known as the "criminal orders"[31] that is so essential to our understanding of the perversion of law and discipline in the Russian campaign. By legalizing murder, robbery, torture, and destruction, these instructions put the moral basis of martial law, and thereby of military discipline, on its head. The army did not simply pretend not to notice the criminal actions of the regime, it positively ordered its own troops to carry them out, and was distressed when breaches of discipline prevented their more efficient execution. At the same time, in calling their troops to order, the generals reminded them of the image and honor of German arms, which their own orders had done so much to besmirch. Moreover, the very application of these orders had a profoundly perverting effect on military language, camouflaging brutalities behind a series of euphemisms and pseudo-legal terms. Ultimately, the army reverted to the crudest moral code of war, according to which everything which ensured one's survival was permitted (and thus considered moral), and everything even remotely suspect of threatening it must be destroyed (and was by definition immoral).[32] Put differently, the Wehrmacht's legal system adapted itself to the so-called Nazi *Weltanschauung*, with all its social-darwinist, nihilist, expansionist, anti-Bolshevik, and racist attributes. This it applied both to its real and perceived enemies, and to its own men.

The close connection between military discipline and the treatment of enemy soldiers and civilians in the Soviet Union can be gauged from the metamorphosis in certain key concepts relevant to both spheres. We have seen that in the West the heaviest sentences were meted to soldiers found guilty of rape, robbery, and plunder. The most highly publicized cases, those which caused the greatest anger among commanders and occasionally culminated in the execution of the culprits, thus concerned attacks on the person and property of enemy civilians. Very few soldiers were charged with serious breaches of discipline concerning their conduct within their units and on the battlefield, for the fighting lasted only a few weeks and victory was quickly and decisively achieved. In the Soviet Union, however, we no longer hear of soldiers being tried, let alone executed, for acts of violence and plunder against Soviet citizens. Indeed, according to the "Barbarossa" decree, such prosecution was legally possible only if it was shown that by committing these offenses a soldier had simultaneously breached military discipline. But officers rarely made use of this argument and preferred to refrain from taking any punitive measures. While the frequency of complaints made by commanders regarding their men's "wild" actions testifies to their widespread prevalence, the almost total absence of courts martial against such offenders shows just as clearly that for all intents and purposes, both *de jure* and *de facto*, they had ceased to feature as punishable breaches of discipline. Conversely, as compared with previous campaigns, soldiers on the Eastern Front became the target of an ever harsher policy of punishment for breaches of discipline related to actual combat activity, as the dramatic rise in long prison terms and executions demonstrates. This was caused by a combination of the conditions at the front and the ideological perception of the war. The tough demands of the fighting made for a growing incidence of attempts to evade the battlefield, while the view of the front as the spearhead of a quasi-religious, anti-Bolshevik, and "racial" crusade meant that such offenders came to be considered as the personal enemies of the Führer and the betrayers of the *Volk*, therefore deserving punishment by death.[33] Because they were fighting against *Untermenschen*, the troops were allowed to treat them with great brutality; but because these same *Untermenschen* threatened Germany, indeed the whole of Western culture and civilization, with a diabolical invasion, refusing to confront them relegated one to their own level. These two spheres of the troops' conduct in the East were also physically and psychologically connected. Under permanent threat of draconian punishment by his superiors if he shrank away from the lethal realities of the front, the individual soldier's compensation was his ability to wield the same destructive power against enemy civilians and POWs. To his officers, he was expendable the moment he ceased to fulfil his functions; to the population, he was the embodiment of the *Herrenrasse*, standing above the law, deciding about death and life according to the dictates of his whim.[34]

The anxiety expressed by several generals on the eve of "Barbarossa" regarding the detrimental effect such murder instructions as the "commissar order" might have on the troops' discipline, though it had no substantial effect either on the formulation or on the implementation of the orders, was based on the experience that once commanders let go of the disciplinary reins, it becomes extremely difficult to regain control over men progressively brutalized by combat.[35] Yet atrocities have often also had a powerful unifying effect on the perpetrators. Let us consider the ancient custom of sacking cities as the culmination of a long siege operation. In this manner commanders compensated their troops both materially and psychologically for submitting to the discipline which had made victory possible, proved to them what profit could be gained by obedience, and allowed them to release their pent up anger and frustration against the enemy, rather than against their superiors. The choice to allow, for a determined spell of time, such localized and controlled chaos, and to direct it toward defenseless members of the enemy's side, instead of leading to the army's disintegration, actually enhanced its cohesion. Something of this mechanism operated in Russia, though on a far greater scale.[36] To be sure, once ordered to take part in "organized" requisitions, it became exceedingly hard to bar the troops from "wild" looting on their own initiative; similarly, having been instructed to shoot certain categories of POWs and civilians, not only did the troops go on with the killing even after the orders were changed, they also took to indiscriminate shooting without regard for the particular categories singled out for murder by their superiors. However, contrary to the expectations of some generals, it was precisely *because*, rather than in spite of what they called the *"Verwilderung"* of the troops, that it became possible to enforce such brutal combat discipline on them without stirring any visible spirit of rebellion, let alone actual mutiny. On one level, it was easier to bear the officers' brutality by being allowed to act brutally toward others; on another, brutal enforcement of will came to be seen as the norm; and, at the most profound level, this vicious circle of brutality merely seemed to confirm the Nazi view of the war—a war whose character had been legitimized by ideological arguments to start with—and thus served to instill into the troops an ever firmer belief in the absolute necessity of fighting and winning Hitler's *Weltanschauungskrieg*. The *Ostheer* was consequently held together by a combination of harsh combat discipline and a general license to barbarism toward the enemy. Though probably not consciously planned, this mechanism produced both impressive combat performance and unprecedented destruction. Most important, these two major aspects of the Wehrmacht's war in Russia were closely tied to and governed by an ideological framework which dictated the army's policies and molded the troops' perception of reality.

The German invasion of Russia, intended to create a vast new *Lebensraum* for the Aryan "race," was launched not only as a war of extermination, but also as an unprecedented campaign of enslavement and plunder. Consequently, the civilian and military authorities planned ruthlessly to exploit the economic resources of the occupied territories, with an eye to satisfying most of the invading army's needs, as well as preserving the German population in the Reich from any war-related shortages. That implementing these policies would bring death by starvation to millions of Russians was fully acknowledged and considered by many as anything but an unwelcome side-effect. Closely integrated into these plans, the *Ostheer*'s combat formations were ordered to "live off the land" with scant regard for the inhabitants' welfare. Such instructions, issued at the start of a ferocious campaign, were bound to have disastrous consequences for the occupied Soviet population. Moreover, as some commanders were quick to realize, the Wehrmacht's policy left Russia's civilians with little choice but to resist with ever greater tenacity an invader who promised them only suffering and death. This had the effect of a self-fulfilling prophesy, for the war in the East soon became precisely the kind of savage struggle for survival Hitler had said it would be. The hardship of fighting called for brutal enforcement of discipline, the intensification of partisan activity led to increasingly barbarous and indiscriminate retaliation by the army. This accelerating process of radicalization, visible at all levels of the *Ostheer*, reflected the true essence not only of the war in the East, but also of the army as a whole, for in the Soviet Union the Wehrmacht finally became Hitler's army in every sense of the term.[37]

The plans of the Reich's Ministry of Food and the Wehrmacht Office of Military Economics and Armaments (*Wehrwirtschafts- und Rüstungsamt*) to turn the Soviet Union into Germany's agricultural, and to a more limited extent industrial hinterland, also had a highly useful propagandistic purpose. Now the regime could present the campaign in the East not only in negative terms as a defensive struggle against the threat of a racially and ideologically demonic enemy, but also as a means to settle all of Germany's domestic problems. This was a war

> for grain and bread . . . a war to create the preconditions for solving the social question. . . . For once we too want . . . to be paid. . . . On the endless fields of the East surge waves of wheat, enough and

more than enough to feed our people and the whole of Europe. . . .
This is our war aim.

Nor did these ideas appeal only to the economic elite of the Reich.[38]
Martin Linder, a twenty-five-year-old chemistry student from Vi-
enna, wrote from the Eastern Front on 22 July 1942 that he expected
the Wehrmacht to take over the whole of Western Russia, all the
way to Astrakhan and including the Caucasus, following which it
would destroy the "Anglo-Saxons" jointly with Japan, and achieve
full control over the Mediterranean:

> Europe will then have more peace, raw materials and time to prepare
> itself for a cleansing of the East. The East will secure for us the
> freedom of food, quite apart from the oil, coal, and iron which will
> come from there in large amounts.[39]

In fact, however, once the invasion began, it soon turned out that
these hopes had been based on a great deal of ignorance as to the
true economic capacities of the Soviet Union, while lack of coordination between the various occupation agencies, along with the
characteristic tendency among both soldiers and civilian adminis-
trators to opt for unrestrained exploitation and destruction rather
than for reconstruction, severely limited those economic benefits
which might have otherwise been reaped. The nemesis of "Barbarossa" was that the very ruthlessness of its exploitive drive hampered
the only kind of economic policy which might have made victory
possible.[40] To this one should add the Red Army's wide-ranging
evacuation particularly of industrial machinery and plants and the
intentional or accidental devastation caused by the fighting itself.
Meanwhile, the needs of the *Ostheer* proved much greater than
expected, especially as the campaign dragged on into winter; and
the extraordinary logistical difficulties often made it impossible to
transport goods from the richer regions of occupied Russia to the
more impoverished areas. Initially the units on the ground were
quite indifferent to the assumption that "in Russia a few millions
can certainly be allowed to die of hunger" if one wished to proceed
with "the ruthless exploitation of the land." But by autumn many
soldiers grasped that this economic policy was turning the population against them. Similarly, some civilian officials began to wonder whether the Germans were not in fact working against their
own ends. As one of them wrote:

If we shoot the Jews, let the prisoners of war die, cause significant portions of the metropolis population to starve to death, and in the coming years also lose some of the agricultural population through hunger, one question remains unanswered: who then is supposed to produce economic goods?

Indeed, by then hundreds of thousands of POWs and civilians were dying of hunger, while others fled to countryside in search of food or joined the partisans. Only in early 1942 was it finally decided to put an end to the brutal and senseless exploitation of the land and systematically to reconstruct its economy, whose collapse merely meant that there would be nothing left to exploit, quite apart from turning the inhabitants against the Germans. Yet the ideological basis of the occupation prevented any far-reaching practical changes in economic policy even then. This was clearly manifested, for instance, by Göring's reaction to reports on food shortages in the Reich, which immediately prompted him to demand from the *Reichskommissare* and military commanders in the East to proceed with the utmost exploitation of the land:

> I . . . do not care if you say that your people [the Russians] are dying of hunger. So they may, as long as not a single German dies of hunger.

Simultaneously, in an attempt to free more of its workers for military service, the Reich turned to importing hundreds of thousands of forced laborers from Russia, once again blocking the possibility of any substantial reconstruction in the occupied areas, and convincing even more inhabitants that they had no choice but to join the partisans, a major cause for the growth of resistance to the Germans also in Western Europe.[41] The result of all this was that by 1943 the Germans produced only 10 percent of occupied western Russia's prewar industrial output, and half of its agricultural yields; by the time even these levels were reached, the *Ostheer* was already on the retreat, laying waste everything it had managed to rebuild in a series of brutal "scorched earth" operations.[42]

The major part of this economic exploitation and devastation was carried out under orders and in a disciplined manner. Nevertheless, the cruelty and brutality such actions involved led many troops to assume that there was no reason to wait for orders from their commanders before they acted in the same manner on their own initiative. For reasons of control and discipline, the officers tried to draw a clear distinction between what they called "organized" and "wild" requisitions, but with singular lack of success. And, as far as the Russians were concerned, the only difference was in any case that organized plunder was generally more efficient, widespread, and ruinous. This was evident from the very first days of the campaign. Upon setting out on "Barbarossa," the 18th Panzer Division ordered its units to rely for their provisions on a "full exploitation of the land." Special "booty-registration-units" were set up and the division expected to need only supplies of flour from the rear.[43] Yet by late July it was realized that the fighting and Soviet evacuations had brought work on the fields to an almost total standstill, and that if it were not resumed immediately, the army would soon be faced with tremendous logistical difficulties, while the population would be doomed to widespread starvation. Consequently the 18th Panzer appointed an "Agricultural Officer" to supervise the reconstruction of the collective farms occupied by its units, and his first report in early August predictably noted that the situation was indeed very serious.[44] Nevertheless, although fully aware now of the potentially disastrous long-term effects of its policies, the division persisted to live off the land without any regard for the future. Official requisitions in August and September 1941 alone amounted to twenty-five tons of meat, rising to forty tons in November, and that despite numerous reports that due to the poverty of the land any further exploitation was out of the question.[45] Similarly, the 12th Infantry Division officially requisitioned no less than 112 tons of oats, 760 tons of hay, 32 heads of cattle, 65 sheep, 94 pigs, 2 tons of potatoes, 350 kilograms of butter, 2350 eggs and 2200 litres of milk between 24 July and 31 August 1941 alone. Such "organized" exploitation was greatly expedited by the centralized Soviet Kolkhoz system. Although the *Ostheer* had made the abolition of these hated collective farms an important element of its propaganda among the farming population of the Soviet Union, it was in fact in no hurry to fulfil these promises in view of the potential of large-scale requisitions and control of production presented by the Kolkhoz. Instead, the army merely replaced managers suspect of "Bolshevik" loyalties by often highly incompetent collaborators. No wonder that by mid-August the 12th Infantry reported that the peasants were completely at a loss (*völlig ratlos*) as to what was actually expected of them by their German occupiers. To make things even worse, the army now ruled that only working men and women would be entitled to a share of the crops, leaving children, pregnant women, the sick, and the old dependent on the good will and increasingly meager resources of the able-bodied.[46] Indeed, by early November the 12th Infantry pointed out that existing food supplies in the areas under its control would "hardly suffice to keep the population alive,"[47] and on 7 December the divisional quarter-master of the 18th Panzer noted that orders for requisitions had taken on an "increasingly theoretical meaning," for the inhabitants had been stripped bare.[48] This armored division had meanwhile also expropriated large numbers of sledges, horses, and snowshoes, all vital for survival in these village communities.[49] But the poverty of the population merely induced these formations officially to sanction "wild" requisitions by instructing their men to resort to "self-help."[50]

The troops had in fact made it their habit to loot the population long before they received official permission to do so. As early as 11 July 1941, at a time when there was still no conceivable existential need for such actions, XLVII Panzer Corps noted that

> the wild requisitions of cattle and poultry . . . from the impoverished inhabitants cause extraordinary bitterness among the villagers . . . requisitions are a matter for the supply officers.[51]

A week later the 18th Panzer noted the high incidence of "senseless" slaughtering of cattle, and by the end of the month it remarked that its troops were often taking the inhabitants' last remaining food reserves and livestock.[52] Yet although the divisional commander rightly warned that such reckless plunder would drive the population "back into the Bolshevik camp,"[53] no one seems to have been prosecuted for this unauthorized looting. In September the divisional commander once more explained that whereas the

> troops should live extensively off the land . . . [this does not mean] that individual units and individual members of the Wehrmacht should try to appropriate supplies on their own initiative [*auf eigene Faust*].[54]

In October II Corps also appealed to the troops not to confuse "necessary and just" requisitions with "wild robbery and plunder."[55] The consequences of the army's policy of exploitation, and of its inability to control and reluctance to punish its soldiers for acts of plunder, were such that in November all front-line formations were informed:

> The livestock population in the occupied parts of Russia has already been so frightfully reduced, that if the unsparing taking of cattle from the land by the troops continues . . . it will result in starvation among the inhabitants and cause severe problems for the German army due to the approach of winter.[56]

But the troops were well aware that official policies were just as ruthless as their own more limited actions and, even more important, that there was little likelihood of being punished. The official distinction between organized and "wild" requisitions was finally shown to have been a farce all along when commanders ordered their troops to loot the inhabitants at will. From now on discipline meant that everything was allowed which promoted the army's survival without thereby threatening its cohesion. The 12th Infantry decreed that "the civilian population is to be allowed to keep what is deemed by the commanders as absolutely necessary for its needs," leaving the interpretation of what the Russians need in order to survive the winter open to junior officers clearly more interested in the fate of their own men.[57] Soon even this reservation disappeared, and units were instructed simply to take everything they needed from the inhabitants. In January 1942, for instance, an order was issued for "felt-boots [to] be ruthlessly taken off the civilian population."[58] One regiment reported the appropriation of forty-eight horses from the villages under its control, leaving the inhabitants with merely two animals, only one of which was capable of pulling a wagon.[59] In the 18th Panzer soldiers were now openly practicing armed robbery, meeting any resistance to plunder with immediate resort to firearms, another habit that once acquired was there to stay, as shown by its reappearance during the following winter.[60] As late as summer 1943, the commander of II Corps complained that

> in spite of numerous and repeated orders forbidding wild requisitions and exactions, such incidents have been reported to the corps time and again . . . [Consequently, the Russians will] understandably view the German soldier as a thief and a robber.[61]

It is quite astonishing that even at this late stage of the war the corps commander failed to grasp that the "wild" behavior of his troops was merely a product of the organized exploitation of the land; indeed, that what had made the individual German soldier into "a thief and a robber" was the fact that his commanders at the highest echelons of the military hierarchy had sent him into the Soviet Union on a campaign of robbery, destruction, and murder.

The effects of these policies on the local inhabitants were predictably disastrous. As the 12th Infantry itself reported regarding winter 1941–42,

> the land was exploited to the utmost. . . . Thereby a situation of general lack of food supplies for the civilian population arose, which in some cases caused starving Russian civilians to turn to German units and ask for relief or beg to be shot.[62]

It is documents of this kind which one should keep in mind when reading the apologetic memoirs of Wehrmacht generals, according to which the "benevolent military administration" was disrupted by the "so-called 'Reich commissars' [who] soon managed to alienate all sympathy from the Germans and thus to prepare the ground for all the horrors of partisan warfare."[63] Not only were the units at the front anything but "benevolent," they also persisted in these exploitive policies notwithstanding reports by their own officers on the horrific consequences of such actions. Even in the first three months of the year the 18th Panzer requisitioned no less than 610 heads of cattle from villages described by its own units as suffering from an acute lack of food.[64] And although it made some attempts to reorganize the agricultural system in spring 1942, the division contradicted its own actions by conscripting all civilians aged over 15 for military-related labor, conducting extensive food requisitions, and, as the second winter approached, expropriating such quantities of horses, sledges, and winter clothing, that according to German units on the ground the population was left without any winter equipment whatsoever.[65] The Grossdeutschland Division pursued the same policies in the Ukraine during the *Ostheer*'s 1942 summer offensive, relying for its food supplies mainly on the local population and, although there were no shortages, allowing its troops to indulge in looting the inhabitants.[66] In autumn the division organized large-scale requisitions of horses, wagons, sledges, and agricultural products, and often also ejected the inhabitants from their houses.[67] With the arrival of winter the troops were urged to resort to "self-help," while unit commander were ordered "to conscript the civilian population with ruthless energy for all tasks" at the front.[68] In the course of this recruitment the division also took the opportunity to point out Jews and other "suspects" found in the villages to the SD for further "treatment."[69] And once more in spring 1943 the division stressed that "no effort should be spared in fully exploiting the economic resources" of the inhabitants, no matter how impoverished they had become in the course of the occupation.[70] The 12th Infantry's growing lack of manpower also induced it to exploit the Russians under its control as forced labor. In January 1942 the division ordered its units "ruthlessly and mercilessly" to recruit civilians for the construction of fortifications, and to "bring the population even to the front-line itself, regardless of age and sex."[71] The following month it was ordered that for "work on the roads, the civilian population is to be conscripted still much more extensively than hitherto."[72] By June 1942 the 12th Infantry was employing 6265 civilians, but maintained that it could only feed 3792 of them; yet another 2208 people, including 476 women and 219 children, were defined as having "no value for the corps and the economy." These so-called "useless" inhabitants were simply driven out of their homes to unknown destinations.[73] Such organized "evacuations" were carried out by orders of the division throughout 1942. Those allowed to remain were mostly employed by the units: the men as *"Hiwis,"* the women and children as forced laborers put to work constructing roads, clearing snow, and building fortifications. In October 1942, for instance, the 12th Infantry engaged 935 civilians in such tasks, of whom 806 were women and children.[74] In 1943 the 18th Panzer also ordered it units "ruthlessly to conscript . . . all the population," women and children included, "with all available means," for work on its defensive lines.[75] The 12th Infantry employed other "young childless girls and women" in various domestic chores within the camps, and there is evidence to suggest that they were used for sexual purposes as well, once more with the approval of local commanders.[76] The civilians employed by the army were officially entitled to merely half the weekly rations supplied to the soldiers, but often received even less than that; nor did they have any medical care whatsoever.[77] The only remedy offered for the frequent outbreaks of epidemics in summer 1942 were "ruthless evacuation" or "complete isolation" of the sick.[78] Meanwhile, an increasing number of civilians were being recruited for work in the Reich. The 12th Infantry sent to Germany 2556 men, women, and children between July and November 1942 alone, and many more in 1943,[79] and the 18th Panzer reported similarly large numbers in spring and autumn 1942. As parents were not allowed to take with them children aged under 15 years, this also created a large population of children whose chances of survival were very slim indeed.[80] Yet all this time the exploitation of the collective farms continued unabated.[81] No wonder that by autumn and winter 1942 the 12th Infantry reported that the population's food reserves were once more "infinitely scarce."[82]

Withdrawals, which especially in the case of Army Group Center were carried out on a significant scale already in winter 1941–42, were another occasion for that deadly combination of official destruction and individual brutality. Reeling under the weight of the first Soviet counter-offensive, the 18th Panzer burned all the villages it was forced to evacuate, destroyed or consumed their entire livestock, arrested and sent to the rear their adult male population, and drove the women and children out into the snow.[83] This was common practice on other sectors of the front as well. Thus, for instance, Werner Pott wrote from the area of Kalinin on 19 December 1941 of "the civilian population, whose houses we set on fire in our retreat and who will be abandoned to death by starvation."[84] Upon reaching its new defensive line, the division rapidly created a so-called "desert zone" some ten miles deep, by driving out the population and laying waste its settlements. Thus on 1 January 1942 no less than forty-eight villages were ordered evacuated and destroyed.[85] Barely a month later the division retreated to yet another defensive line, and once more created a stretch of barren territory ahead of its positions, burning down houses, poisoning wells with dead cattle, taking the men to the rear, and ordering the women and children "to wander off to the area north-west of the desert-zone" in temperatures reaching 40° Celsius below freezing point.[86] Similar tactics were employed during the division's withdrawals in winter and summer 1943, and along with the special *"Kommando"* established especially for the purpose of destroying all the economic assets in the areas vacated by the formation, and the arrest of all adult men and women, the troops looted whatever they could get their hands on.[87] It was the same case during the withdrawal from the Demyansk enclave in winter 1943. The formations carefully planned and methodically executed the complete devastation of the region by destroying and booby-trapping the villages, slaughtering the cattle, and burning down agricultural installations and machinery; to deprive the Red Army from any potential conscripts, the men were arrested and sent to the rear.[88] This "scorched earth" operation was carried out with such discipline and efficiency that II Corps publicly praised its formations for their excellent achievements.[89] The GD's role in such actions in the course of 1943 was particularly conspicuous, presenting its troops with a wide range of opportunities for individual brutality.[90] During the retreat to the Dniepr in autumn 1943, the division established a so-called *Räumungskommando*, which within the space of three weeks drove out of their villages no less than 13,627 civilians, requisitioned 9268 heads of cattle and 1392 tons of crops, and destroyed 1260 agricultural machines and 165 mills. However, as the final report of this unit noted, these figures did not include the widespread "wild" looting triggered by the official measures.[91]

Of arguably even deadlier consequences was the manner in which the Wehrmacht's orders regarding Soviet soldiers and politically or "racially" dangerous elements not only officially sanctioned a campaign of organized murder, but also opened the way for a massive wave of indiscriminate shooting by soldiers who refused to distinguish between the various categories of enemies dictated from above. Here was indeed a powerful demonstration that by perverting the moral basis of discipline the army had undermined its own ability to enforce such orders which of necessity had to apply moral arguments. By providing its troops with a license to murder disarmed soldiers and defenseless civilians, the army could no longer punish them for persisting with such actions when instructed to stop, nor to insist on making fine distinctions between the victims. Contrary to postwar claims, while orders to kill were carried out with considerable efficiency, orders to confine the murders to certain categories of people and then further to limit them due to changing circumstances were widely ignored by the troops. Significantly, though in both cases refusal to follow orders was rarely punished, indiscipline in the former was just as rare as it was frequent in the latter.

In the course of the Russian campaign over 5,700,000 Red Army soldiers were captured by the Germans, of whom no less than 3,300,000, or 57 percent, died. Indeed, even by early 1942 two million Soviet POWs were already dead. This unprecedented death rate was related to the execution of commissars by the troops upon capture; to the delivery to the *Einsatzgruppen* for "special treatment" of so called "politically intolerable" (*politisch untragbaren*) prisoners, that is, all members of the intelligentsia (*Intelligenzler*), "fanatic communists," and Jews; and to explicit orders issued to the formations on the ground to supply POWs "only with the most essential provisions," and to "feed them with the most primitive means," as well as to the lack of any serious preparations for the huge number of prisoners the Wehrmacht expected to take by employing its well-tried envelopment tactics against the Red Army. Thus while as many as 600,000 prisoners may have been shot outright, many of those who survived the initial selection were doomed to perish not long thereafter from starvation, exposure, epidemics, and exhaustion. Commanders at the front repeatedly warned their men of the "treacherous behavior especially of prisoners of war of Asian descent," ordered them to "act ruthlessly and energetically against the slightest sign of insubordination" and "totally to eliminate any active or passive resistance" by making "*immediate* use of weapons," and reminded them to "take into account the animosity and inhuman brutality of the Russians." Not surprisingly, these instructions led to an upsurge in indiscriminate shootings. Thus a further unknown, but probably very large number of Soviet soldiers were murdered by the troops following their surrender but before they were given the opportunity to be counted as prisoners. To this killing of POWs was added the "destruction" of so-called political and biological enemies, mostly simply described as "bandits" or "partisans," without much effort to distinguish between real guerrillas, political "suspects," and Jews. The extent to which this euphemism was applied to what were in fact large-scale murder operations was demonstrated, for instance, by the report of the Wehrmacht commandant of Belorussia, who claimed to have shot 10,431 prisoners out of 10,940 taken in "battles with partisans" in

October alone, all at the price of two German dead. Yet this was but one of many so-called "anti-partisan campaigns" which turned out to be outright massacres of unarmed civilians.[92]

Bearing in mind the frequently heard argument that combat formations had had no part in the execution of the "criminal orders," the extent of the official and unauthorized actions taken by the troops at the front is particularly striking. On the eve of "Barbarossa" the 12th Infantry Division ordered its units to separate officers, commissars, and NCOs from all other prospective prisoners and deliver them for further "treatment" to the SD.[93] Indeed, divisional interrogation records show that long after the *Kommissarbefehl* had been officially rescinded, prisoners were still being sorted out along "racial" criteria, thereby greatly facilitating their "elimination" by the death squads.[94] It must also be stressed that both the generals in the rear, and even more so the officers on the ground, tended to choose the most radical interpretation of the "criminal orders." For instance, in mid-September OKH added the provision in its orders to combat divisions in the East that all Soviet troops who had been overrun by the Wehrmacht and had then reorganized behind the front were to be treated as partisans (that is, to be shot on the spot).[95] This instruction was more severe than the "Barbarossa" decree, but still left some leeway by specifically referring to organized soldiers and employing a euphemism. The 12th Infantry's commander, however, had no time for such subtleties, and summarized his instructions at an officers' conference in the following manner:

> Prisoners behind the front-line. . . . Shoot as a general principal! Every soldier shoots any Russian found behind the front-line who has not been taken prisoner in battle.[96]

Yet even before these orders were issued, reports of indiscriminate shootings by the troops rapidly multiplied. As early as 5 July 16 Army warned its formations that "after POWs have been reorganized into work-battalions they should not be attacked and shot,"[97] a clear indication that even prisoners forced to volunteer their services to the army were not safe from trigger-happy German soldiers. Two weeks later the 12th Infantry urged its troops to understand that " 'bumping off' ['*Umlegen*'] Russians who had already been taken prisoner is unworthy of German soldiers."[98] But neither in this division, nor in many others, were the troops convinced by such appeals, particularly as they were normally accompanied by orders to kill a great many other Russians belonging to a growing list of political, "racial," and military categories. Indeed, as the army's propaganda represented all Russians as *Untermenschen* not deserving of life in any case, the soldiers saw no reason to distinguish between them and those slated to be shot outright.

The armored formations of Army Group Center are a good case in point. Here the official policy of selective murder outlined by the "Barbarossa" decree had an immediate and striking effect on the troops. On 25 June 1941, that is, on the third day of the campaign, the commander of XLVII Panzer Corps issued the following order:

> I have observed that senseless shootings of both POWs and civilians have taken place. A Russian soldier who has been taken prisoner while wearing a uniform, and after he had put up a brave fight, has the right to decent treatment.

From this passage one would assume that General Lemelsen considered the Russian campaign to be a war like any other, and for this reason was shocked by his men's behavior. Yet as he proceeded to explain to the troops the real goals of the campaign, the corps commander revealed that his own thinking had been molded by that combination of ideology and ruthless, indeed cynical practicality typical of National Socialism: "We want to free the civilian population from the yoke of Bolshevism and we need their labor force." The Russians were to be freed from Bolshevism so that the Germans could enslave them for their own purposes. At the same time, the political goal of destroying Bolshevism was to be carried to its ultimate conclusion. Hence the general hastened to clarify that his instructions to spare the lives of the prisoners did not apply to their political leaders:

> This instruction does not change anything regarding the Führer's order on the ruthless action to be taken against partisans and Bolshevik commissars.[99]

This meant that both the political officers in uniform, and anyone described as a partisan by the army, a category which included "racially undesirable elements" such as Jews, were to be done away with. And, while Red Army soldiers belonging to neither criteria were to be initially spared, they too could expect to perish later or at best to become the Reich's slaves. Under such circumstances, it is no wonder that merely five days later Lemelsen had to appeal once more to the troops, as they had entirely ignored his first order. Yet significantly, even now the general did not threaten to punish the culprits, in spite of the fact that he was confronting what amounted to a collective breach of discipline:

> In spite of my instruction of 25.6.41 . . . still more shootings of POWs and deserters have been observed, conducted in an irresponsible, senseless and criminal manner. This is murder!

But again the corps commander felt obliged to explain to his troops what the war was about, thereby belying his overt intention of conducting the fighting according to the traditional rules of war:

> We want to bring back peace, calm and order to this land which has suffered terribly for many years from the oppression of a Jewish and criminal group.

So as to make sure that the men did not misunderstand his criticism, nor take his ideological clarifications as mere theoretical rationalizations, the general repeated in even greater detail the murder instructions which had set off the brutalities he was trying to stop:

> The Führer's instruction calls for ruthless action against Bolshevism (political commissars) and any kind of partisans! People who have been clearly identified as such should be taken aside and shot only by an order of an officer.[100]

So much for Guderian's claim that his Panzer Group 2, one of whose elements was XLVII Panzer Corps, never even received the "commissar order."[101] More important, as this incident clearly revealed, soldiers could and did disobey orders, and though they were occasionally admonished for doing so, as long as their disobedience concerned greater brutality against the enemy, rather than attempts to evade combat, they were hardly ever punished. It was now obvious that attempts to control the troops while issuing them with a license for murder could not possibly succeed. It should also be understood that Lemelsen was representative of many other *Ostheer* generals, who though appalled by their men's brutality, were simultaneously engaged in furnishing them with arguments lifted directly from Hitler's ideological arsenal as a means to motivate them in battle and make them believe that the murders they were ordered to carry out were an unavoidable existential and moral necessity.

This pattern was repeated over and over again during the war in the East. Whereas the recognition by the generals that the killing and maltreatment of POWs merely stiffened the enemy's resistance, coupled with the need for forced labor in the Reich, finally caused the abolition of the *Kommissarbefehl* and brought about some improvement in conditions for prisoners by 1942, commanders failed entirely in their efforts to put a stop to the indiscriminate shootings by their troops. Ironically, in this respect breaches of discipline in the *Ostheer* were far more common than in any other army in the Second World War; even the Red Army was more successful in controlling its men once they entered German territory than the Wehrmacht had been in Russia. Under the pressure of combat and ideology, harsh discipline and official barbarism, the men were progressively brutalized. At this stage one could no longer expect them to alter their conduct toward an enemy still described as a devilish *Untermensch*. The contradictory practical interests of formations on the ground also constantly clashed, for while commanders were highly conscious of the detrimental effects of their troops' conduct, they countered their own efforts to control them by simultaneously attempting to stiffen the men's resolve by means of ideological indoctrination. In February 1942 the 18th Panzer admitted that "Red Army soldiers . . . are more afraid of falling prisoner than of the possibility of dying on the battlefield," and maintained that this was manifested by the fact that "since November last year . . . only a few deserters have come over to us and that during the battles fierce resistance was put up and only a few POWs were taken." The division concluded that "the troops must be instructed that not all Russians are communists."[102] However, it was precisely at this time of military crisis that this same division, along with the rest of the *Ostheer*, greatly intensified the ideological indoctrination of the men with the hope of thereby making up for its tremendous material attrition.[103] Nor was the indiscriminate shooting of prisoners limited to soldiers who had grown used to operating according to the dictates of the "commissar order." The GD Division, which came to the front after the order was rescinded, behaved in precisely the same manner, demonstrating that this was more a question of ideological preparation and unwillingness on the part of commanders to enforce their will as regards Russian lives. In September 1942, for instance, the GD instructed its troops that

> all commissars—politruks—who fall alive into the hands of the troops are to be transferred immediately to the divisional intelligence section. Shooting by the troops after taking them prisoner is strictly forbidden.[104]

As we now know, army intelligence collaborated closely with the SD, and there is good reason to believe that "suspects" were eventually shot. But the troops, not threatened with any punishment, had no time for these somewhat more cumbersome procedures. Far from being ignored while still in force, the "commissar order" kept dictating actions on the ground long after it was officially withdrawn, especially as the other criminal sections of the "Barbarossa" decree remained in operation throughout the war. Unable, and probably also unwilling, to prosecute their men for killing Russians and Jews, front-line divisions were satisfied with simply recording the innumerable cases of indiscriminate shootings throughout the war. In another such instance the GD noted that

> many deserters brought wounded [from the battlefield] have claimed that they had received their wounds from the Germans hours after they had thrown down their weapons and had visibly indicated their intention to desert by raising their arms in the air.[105]

And in April 1943 the division issued yet another rather limp appeal to its troops, emphasizing that they

> must understand that the ultimate result of the maltreatment or shooting of POWs after they had given themselves up in battle would be . . . a stiffening of the enemy's resistance, because every Red Army soldier fears German captivity.[106]

All this had little effect for, as their experience had taught them, the troops knew that orders not backed up by a willingness to resort to harsh punishment were not really meant to be obeyed in the first place.

Calls by formation commanders to behave toward POWs with more consideration must have been viewed as quite cynical at least by some of the soldiers who remembered the official maltreatment of prisoners particularly during the first months of the campaign. 16 Army, for instance, ordered its formations on 31 July 1941 not to transport POWs to the rear in empty goods and troop trains as they might "contaminate and soil" them.[107] Thus the ill-fed prisoners were forced to walk across vast distances to camps in Poland, with the result that many of them died on the way. The 18th Panzer similarly forbade the transportation of its POWS in trucks, for fear they might infest them with lice.[108] Prisoners employed as forced labor by front-line divisions were officially entitled to just over half the rations given to German soldiers, but even this insufficient amount was drastically reduced in late October.[109] In practice, many formations, such as the 18th Panzer, insisted that prisoners be fed "with the most primitive means."[110] Especially in winter, combat divisions also ordered their troops to confiscate from the prisoners all items of clothes deemed useful for their own protection from the cold, condemning POWs to death by exposure.[111] Medical treatment of wounded prisoners by German doctors, as well as providing them with any army issue medicines, was also officially forbidden.[112] Yet as the manpower shortage worsened, growing numbers of POWs were "nominated as volunteers" and employed in a variety of military tasks, among them clearing minefields without receiving any previous training.[113] These official actions set an example to the troops as to the manner in which Soviet prisoners should be treated, and no amount of appeals later on in the war could undo their impact.

If the *Ostheer*'s commanders found themselves in some conflict with their troops over the treatment of POWs, there was much less friction regarding the so-called partisans, though here too the soldiers found ample opportunities for "wild," unauthorized actions. "Partisans," or "bandits," was a term used to describe all civilians deemed unworthy of life by the army, whether due to guerrilla activity or to political and "racial" affiliation. In a world where life was cheaper than food and clothes, and whose inhabitants were divided into members of the "master race" and "subhumans," one could not issue orders for massacres of helpless civilians without expecting the perpetrators to perform similar atrocities on their own initiative. Sanctioned already in the "Barbarossa" decree, collective punishment for partisan attacks soon became the rule, and from that point on no one in the army seemed to bother much about the identity of the murdered. On the first day of the campaign the 12th Infantry instructed its units not to treat guerrillas (*Freischärler*) as POWs, but rather to have them "sentenced by an officer on the spot,"[114] one more euphemism for summary execution. In late July 16 Army similarly ordered that members of "partisan-battalions," as well as all civilians rendering them any kind of assistance, were to be "treated as guerrillas," that is, to be shot.[115] The 18th Panzer issued very similar instructions in early August, ordering its troops to shoot anyone resisting its units and not clearly identifiable as belonging to an organized army unit, as well as any civilian suspected of having rendered assistance to resisters.[116] The division also enforced strict curfew regulations on occupied settlements and instructed the troops to make "ruthless use of firearms" against offenders,[117] almost precisely the same order given by II Corps, which insisted that "Ruthless action is to be immediately taken against suspect elements!"[118] This too implied either shooting or, as was becoming increasingly popular in the *Ostheer*, hanging, for "suspect elements" meant anyone belonging to "undesirable" political and "racial" categories.[119] Indeed, as early as 4 July the 12th Infantry reported the execution of ten civilians accused of belonging to the communist youth organization or of being Jewish.[120] Although it was soon discovered that this policy only enhanced guerrilla resistance,[121] in mid-August II Corps insisted that "partisans are to be publicly hanged and to be left to hang for some time" as a warning, while communist bureaucrats "about whom nothing can be proved" should be delivered to the *Einsatzgruppen*, obviously for somewhat more discreet execution.[122] The following month the 12th Infantry declared that anyone seen "walking about" during curfew should be shot, and that village mayors would forfeit their own and their families' lives for any partisan activity in the vicinity of their villages,[123] a threat also made by the 18th Panzer in October, accompanied by the taking of hostages to prove its serious intentions.[124] By then the 12th Infantry decreed that anyone "tolerating" partisans would be hanged, and a few days later it was announced that civilians caught without the recently issued passes would be shot on the spot.[125] The troops were meanwhile expressly ordered to burn down any house suspect of harboring partisans without any attempt to find out who was actually inside.[126] In November divi-

sional units were detached to carry out reprisal actions against villages which had failed to report the presence of "aliens," and while
such "suspect elements" were either executed by the troops or given
over to the SD for "special treatment,"[127] other units became engaged in a campaign of public hangings of civilians charged with
such offenses as "feeding a Russian soldier," "wandering about,"
"trying to escape," and so forth.[128] Between 11 November and 5
December II Corps "annihilated" 448 "partisans" and burned down
sixteen villages, driving off their cattle and horses and destroying
all existing food reserves. As the corps admitted the loss of merely
six soldiers killed and eight wounded in a partisan attack far from
this area, it is obvious that this frenzied extermination policy had
very little to do with actual guerrilla activity.[129] This pattern was
repeated in another such operation reported by 16 Army, where
387 civilians were killed at the price of ten dead and eleven wounded
soldiers.[130]

As the front stabilized, more systematic steps were taken against
the population. In December the 12th Infantry evacuated some 2000
inhabitants from its front to a depth of six miles, either burning
down their houses or taking them over for accommodation.[131] Even
II Corps conceded that the civilians were driven off "with entirely
inadequate supplies of food," all in the height of winter.[132] Behind
this area, another so-called "barred zone" was established where,
by the 12th Infantry's own account, provisioning the population
became "quite impossible."[133] Nevertheless, inhabitants disobeying
any of the rules imposed on their movement were either shot on
the spot or delivered to the SD.[134] This only forced more people to
join the partisans. But instead of changing what had clearly become
a counter-productive campaign of murder and destruction, in January Army Group North declared that

> The recent revival of partisan activity in the rear area... demands
> that action be taken... with the greatest ruthlessness. Partisans
> should be destroyed wherever they appear, as should their hiding
> places [i.e., villages], if they are not needed by our troops for
> accommodation.[135]

Indeed, by this time the 12th Infantry's standard reaction to partisan
attacks on its units was to burn down the villages in the vicinity of
the attack, shoot all the male inhabitants, and leave the remaining

women and children to fend for themselves in winter conditions without any shelter.[136] The 18th Panzer also practiced a policy of summary executions of suspects behind its "desert zones" throughout 1942.[137] Similarly, on the eve of the Wehrmacht's summer offensive in the Ukraine the GD Division ordered its troops to "destroy" all captured partisans, and gave license to battalion commanders to execute civilians suspected of rendering any assistance to guerrillas.[138] Consequently, by September reports of indiscriminate shootings of civilians multiplied to such a degree that the division decided to transfer the responsibility for executions to the GFP, the Secret Field Police which closely collaborated with the SD.[139] But though the troops were explicitly ordered to transfer all suspects to the police, the killing went on unabated. Combat divisions were also employed in large-scale "anti-partisan" operations. In May 1943 the 18th Panzer participated in *"Zigeunerbaron,"* a "cleansing" action in the forest regions south of Briansk. Troops were ordered to arrest all male civilians between the ages of fifteen and sixty-five, and to drive out the remaining population, whose property was to be confiscated and villages burned down. Captured Red Army officers and commissars were to be transferred to the division's intelligence section, most probably for later execution by the SD, while soldiers, communist party members, and Jews were to be used for clearing minefields, the Wehrmacht's idea of a profitable execution. Indeed, within the space of merely two weeks the division and its support units killed 1584 "guerrillas," took 1568 prisoners, and drove out of the area 15,812 inhabitants, burning down all their villages. The fact that the division found only a few obsolete firearms among the population, and that it sustained almost no casualties, indicated that this operation was merely one more feast of destruction against defenseless civilians.[140]

The extreme measures used against civilian resistance to German occupation, and particularly the implementation of the regime's racial and political elimination policy by the army under cover of "anti-partisan" operations, ordered, organized, and perceived as quite legal by military commanders, had a powerful brutalizing effect on the troops. Very few units on the Eastern Front could have avoided taking part in such actions. And, as what constituted guerrilla activity included a wide-range of actions or lack of actions, and as the term "partisan" denoted not merely active or passive resisters, but also people belonging to "undesirable" political and "racial" categories, the war in the East offered the German soldier endless opportunities for committing authorized and unauthorized acts of murder and destruction, robbery and plunder, rape and torture, for which he was rarely punished and not infrequently praised by his superiors. Nevertheless, when the term "partisan" seemed insufficient to legitimize brutality, especially where obviously helpless civilians were concerned, the army sometimes resorted to the euphemism "spy" or "agent," a uniquely useful term precisely because it was based on the assumption that innocence was the best indication of guilt. In July 1941 the 12th Infantry tried out this euphemism for the first time when it ordered its troops to shoot refugees attempting to cross the German lines "for suspected espionage."[141] The troops reacted with such enthusiasm, and the incidence of indiscriminate shooting rose so dramatically, that II Corps was finally compelled to order "suspects" to be brought before the appropriate officers for interrogation, rather than simply killing them on the spot.[142] The 18th Panzer, which had also treated refugees as an "espionage hazard,"[143] took till June 1942 to admit that "grave errors" had been committed in the treatment of "agents," and even then it merely instructed its units to transfer "suspects" to the GFP, thereby sentencing them in any case to almost certain death.[144]

Now while the generals had little scruples about issuing orders to shoot men and uproot whole populations, they feared that executing women and children might cause disciplinary problems among the troops, and normally preferred the SS and SD to carry out such unsavory tasks.[145] But in the numerous "unavoidable" cases of brutality against women and children, as well as in the army's propaganda against fraternization, the euphemism "agent" was found to be very handy indeed. In October 1941, for instance, the 12th Infantry warned its troops that "information is usually carried by youngsters aged 11 to 14," and recommended "flogging [as] the most advisable measure for interrogation." The Russians, it explained, "and especially the women," were liars, but "a few pats on the back will shorten" the process of interrogation considerably.[146] An example of the senseless sadism such ideas inspired was the interrogation of fifteen "agents" conducted in spring 1942 by II Corps. The "suspects" included twelve lads aged between fifteen and seventeen, a mother with her child, and only one adult male. The corps reported that two of the boys, though manifesting "a certain stupidity," were "fanatic communists and German-haters," and consequently their interrogation "lasted many hours," and "every truthful statement had to be forced out of them by the most brutal methods." Subsequently all "suspects," including the mother and her child, were executed as "agents."[147] The GD Division explained to its troops in October 1942 that "men of all ages, good-looking women and particularly young girls and lads and even children," were being employed as enemy agents; children were supplied with unlikely "stories," such as that they were looking for their parents, while their real object was to spy for the enemy.[148] The 12th Infantry tried to dissuade its men from fraternizing with Russian women by pointing out that they were "mostly Jewish females . . . whose Jewish origin cannot be seen."[149] Along these same lines the 18th Panzer warned that "the Russian woman is prepared to make unscrupulous use of her physical advantages and of our soldiers' trust for purposes of espionage in the interest of the war," and that not only was "intercourse with female civilians . . . unworthy of the German soldier," it "also carries with it the danger of being exploited or harmed by a spy, of falling into the hands of a female partisan and of being terribly mutilated." The division thus threatened the troops that while women caught in the company of German soldiers would be delivered to the police on "security" grounds, their male partners would be prosecuted for collaborating with the enemy's intelligence service.[150] Similarly, the GD reminded the men that whereas "contact" with women was likely to confront them with an enemy "agent," it was certain to end with some kind of venereal disease, as nearly all Russian women were "known" to be infected.[151] This campaign did not necessarily prevent the troops from fraternizing with Russian women, but it may well have prepared the background for their brutal behavior and indifference to the fate of helpless women and children when the occasion arose during "anti-partisan" operations.

The *Ostheer* thus provided its troops with an array of orders, rationalizations, and incentives for brutally treating both captured enemy soldiers and the occupied civilian population. The scope of officially authorized murder, maltreatment, and destruction of property far outweighed "wild" actions performed on the troops' own initiative. Moreover, the troops' conduct can only be understood within the context of the Wehrmacht's far-reaching legalization of actions previously considered criminal, the organized manner of their execution, and the widespread agreement with the ideology which motivated them, all of which made for a situation whereby an army normally insistent on rigid obedience allowed its troops to get away with mass breaches of discipline regarding the treatment of enemy soldiers and civilians. Paradoxically, these "wild" acts of brutality made it possible for the officers to enforce the harsh combat discipline essential for maintaining the cohesion of the army. For their crimes against the enemy the troops paid a high price by yielding to the draconian measures applied by their commanders.

Discipline in the German army was always harsh; but in the Wehrmacht, and especially in the *Ostheer* of 1941–45, it became positively murderous. This had of course to do with the conditions at the front, but more important, it was a manifestation of the extent to which the regime's *Weltanschauung* had penetrated the ranks of the army and remolded its concepts of legality and criminality, morality and justice, discipline and obedience. Indeed, here was a clear case of the connection between the the army's crimes toward other nations and its ruthless treatment of its own troops, for both stemmed from the same ideological roots, and were based on the same inversion of martial law. Being simultaneous, these two aspects of the Russian campaign also had the effect of enhancing and legitimizing each other, that is, of accelerating the process of brutalization both within the ranks of the army and toward the enemy, and creating a need for ever more ideological justifications, which in turn provided arguments for even greater brutality. Hence the army's legalization of crimes toward the enemy, its toleration of disciplinary offenses by the troops toward the same enemy, and its enforcement of a brutal discipline as concerns the soldiers' combat performance, were all linked to each other, derived their legitimacy from the acceptance of the Nazi "world view," and must be seen together as forming the kernel of the war in the East, and by extension, as the most characteristic and essential features of the Wehrmacht.

With this background in mind we should examine the unprecedented record of military justice in the Wehrmacht. Whereas during the Great War the Kaiserheer executed only forty-eight of its soldiers, in the Second World War between 13,000 and 15,000 men were put to death by their own army. By way of comparison, we may note that whereas in 1914–18 the British and French armies executed far more soldiers than the Germans (346 and 650 respectively), in 1939–45 the proportions were drastically reversed, with only forty British and 100 French soldiers being put to death. The astonishing number of executions in the Wehrmacht was mainly due to the politicization of martial law, whereby such offenses as desertion and self-inflicted wounds came under the heading of treason and subversion (*Wehrkraftzerseztung*), and were consequently punishable with death. Thus, between January 1940 and March 1942 some four-fifths of the death sentences were based on ideological-political grounds. It has been calculated that an average of 100 soldiers accused of desertion, and another 100 found guilty of subversion, were executed every month throughout the war, constituting about half the total number of soldiers charged with these offenses. Moreover, by 1944 as many as 85 percent of all death sentences passed were actually carried out, as opposed to only 30 percent in the Great War. As the war dragged on, the number of trials per month rose by a factor of 3.5, from 12,853 in December 1939 to 44,955 in October 1944, but the number of death sentences rose by a factor of no less than eight, from 519 in 1939–40 to 4118 in 1943–44. This meant that in relation to trials, the death sentence became twice as common. Soldiers who escaped capital punishment could still expect extremely severe sentences. Quite apart from the thousands of men sent to penal battalions, between late August 1939 and mid-1944 as many as 23,124 soldiers received very long prison terms with hard labor, 84,346 were sentenced to over one year's imprisonment, and yet another 320,042 men were sentenced to periods of less than one-year's arrest.[152]

The reluctance of commanders in Russia to prosecute their troops for offenses committed against enemy soldiers and civilians was thus more than offset by their quick and brutal retaliation against men charged with breaches of combat discipline. This can be best seen at the level of the fighting units themselves. On the eve of "Barbarossa" the commander of Panzer Group 2 instructed his officers to take the most severe measures against breaches of discipline, and made a point of stressing that deserters "should be shot on the spot."[153] By 1 August 1941 the 18th Panzer, which belonged to this

Panzer group, informed its troops that three soldiers captured by the Russians had committed treason by telling the enemy "that they had been forced to fight," and that consequently, once back in German hands, they could expect no mercy "for such dishonorable behavior."[154] A couple of weeks later the division tried three soldiers for cowardice, sentenced one of them to death and the other two to ten years' imprisonment and the loss of all their rights as citizens. "This incident," the divisional commander briefed his officers,

> is to be used as an urgent theme for instruction. It should be pointed out that cowardice is not only the most disgraceful, but also one of the most dangerous crimes a soldier can commit, for it not only undermines discipline, but also weakens the striking capability of the troops. This danger will be dealt with in every instance with the heaviest punishment—the death penalty.[155]

This was no idle warning. In mid-December, for instance, an NCO who had pulled his squad out of its battle positions due to what he thought were the sounds of approaching Soviet tanks, was sentenced to ten years' imprisonment.[156] At the end of the month the 18th Panzer issued a special order clearly intended to emphasize that the consequences of cowardice should be feared even more than its causes:

> Lance-Corporal Franz Aigner, staff company II/Panzer Regiment 18, was sentenced to death by court martial on the charge of cowardice. . . . Although he had seen his unit marching forward, he entered a house, drank a bottle of schnapps with an unknown soldier, though he had already drunk a sufficient amount of alcohol before, and fled to the rear without cap or weapon, where he was seized in this ragged and drunken condition. Every case of cowardice will be severely atoned for with death. The troops are thoroughly to be instructed on this by the company commanders personally.[157]

As the chaos at the front made for a rise in breaches of combat discipline, lesser offenses were also punished with great harshness. In late November 1941 the 18th Panzer tried eight soldiers for neglect of duty while on guard, sentencing three of them to between three and four years' imprisonment and the remaining five to even longer terms. Another soldier, found sleeping in his position, was sentenced to five years in jail. But even this seemed insufficient, and the divisional commander took up the issue in another appeal to his troops:

> As all instructions and warnings, as well as the publication of previous sentences by court martial to long prison terms have hitherto had no instructive and deterring effect, the court martial is compelled to threaten with making use in the future of the heaviest punishment—the death sentence.[158]

It thus became standard policy to terrorize the troops from evading a likely death at the front by promising them certain execution if they were caught in the act.

In the course of 1942 the soldiers had ample opportunity to learn that their commanders were quite willing to carry out their threats. In March, for instance, two soldiers belonging to the 18th Panzer were sentenced to death, one for abandoning his gun crew during combat and another for refusing an order in the course of an enemy attack.[159] That same month XLVII Panzer Crops noted:

> The discipline of the troops has deteriorated during the winter. This is shown also by the increasing number of trials. Guard duty has particularly been neglected, spelling danger for the troops.

But although it was obvious that these were manifestation of exhaustion due to heavy combat and lack of manpower, the corps insisted:

> The troops are repeatedly to be instructed that only the most severe punishment can be expected in cases of cowardice or fear, desertion, absence without leave and neglect of duty while on guard.

To be sure, in some cases it was precisely the fear of such severe punishment that made soldiers who had absented themselves for a brief time from their units to turn into real deserters. As the corps itself realized:

> There have been numerous cases in which soldiers who had left their units or were absent from them, kept going to the rear only because they were afraid of being shot.

Consequently, the corps promised the troops: "A soldier who reports back to his unit of his own free will, will always receive a lighter sentence than one who is caught." But light sentences were a very relative concept at the time, and the number of severe punishments in fact kept multiplying. Thus soon thereafter the corps executed another deserter, sentenced to death (but ultimately did not execute) a soldier charged with self-inflicted wounds, sentenced two men to six and eight years' imprisonment respectively for smoking during a partisan attack on the village in which they were quartered, and sentenced to three years' imprisonment with hard labor and loss of all citizenship rights a soldier who had returned late from an assignment in the rear.[160] The commander of the 18th Panzer was evidently losing his patience with the troops, as could be seen from the notes he prepared for a conference with his officers:

> Four deserters in the area of the corps—thorough instruction—deserters will never see the Fatherland again—concerning all those reported "missing" conduct a thorough investigation.[161]

The GD Division, plagued by similar difficulties, reacted with no lesser harshness, sentencing to death seven men in the second half of 1942 alone, and charging another eleven soldiers with desertion, most probably sentencing them too to death, though their records have not survived.[162] The 12th Infantry Division, which sentenced to death at least sixteen men charged either with desertion or self-inflicted wounds between the beginning of "Barbarossa" and June 1943, made a special effort to publicize the executions, whose effect was enhanced by posthumously depriving the culprits of their rights as German citizens and thereby barring their families from collecting their pensions.[163] This policy doubtlessly terrorized the troops. As a deserter from the division reported to his Soviet interrogators in October 1942, the reason that in spite of the terrible living conditions, the devastating casualties, and the prevalent sense of pessimism among the troops, discipline still held was mainly fear. The men were afraid that if they deserted their families would be punished, that if they were seen trying to cross over they would be shot, and that if they were caught they would be executed. Indeed, he claimed that he had succeeded in reaching the Russian lines only by shooting three sentries who had opened fire at him.[164] Nor was he exaggerating. A few months earlier 16 Army informed its units of the summary execution of five soldiers caught while trying to desert.[165] A Russian deserter told his interrogators that of seven German soldiers who had tried to cross over to the Red Army, four were shot dead by their own officers before they could make it to the other side.[166] As a further humiliation, soldiers executed for desertion were now buried without military honors, far from any military cemetery, their graves marked with a simple sign indicating only name, date of birth and death, but neither rank nor unit designation.[167]

By the third year of the war in Russia ever harsher measures were needed to keep the men at the front. In February Hitler circulated the following order to all front-line formations:

> I have found out that during the retreats and evacuations ordered in the last few weeks, there have been some unpleasant and unruly scenes. . . . This is unbearable. . . . The reason for this is that commanders do not make use of all [disciplinary] measures. . . . The harder the times, all the tougher should be the measures by which the commander enforces his will. I therefore demand that every commanding officer and NCO, or in extraordinary situations every courageous man, will enforce the execution of orders, if necessary by the force of arms, and will immediately open fire in case of insubordination. This is not only his right, but also his duty.[168]

Indeed, from this stage on, the Wehrmacht experienced a growing incidence of cases whereby discipline was enforced not merely by court martials and heavy punishments, but also by the application of force without any prior legal proceedings. For this reason it is also impossible accurately to estimate the actual number of soldiers executed for real or perceived offenses during the war. Here was another manifestation of the complete collapse of the rule of law, so characteristic of the Third Reich, and the self-willed realization of the Nazi social-darwinian myth of the survival of the fittest. The army, whose cohesion depended on discipline and obedience, chose increasingly to ignore the legal basis, perverted as it in any case had become, of its disciplinary system, bypassed the cumbersome proceedings it necessitated, and, emulating its own long-established practice against the enemy, simply shot anyone who hampered it from fulfilling its tasks. Hitler's instructions were rapidly applied in field formations. During the evacuation of the Demyansk enclave, II Corps established so-called "Army Patrols" (*Heeresstreifen*), and ordered them "to make immediate use of their weapons (submachine guns, pistols) in order to enforce obedience and discipline."[169] Similarly, in summer 1943 the commander of the 18th Panzer ordered both his officers and soldiers to enforce combat discipline if need be by resorting to the use of arms against their comrades:

> I expect every officer, NCO and man, who has retained his soldierly honor, to do everything in order to control such outbreaks of panic [as described in the first part of the order]. . . . I expect officers to make ruthless use of all means at their disposal against men who bring about occurrences of panic and who leave their comrades in the lurch and, if necessary, not to refrain from using their weapons.[170]

Having legalized the murder of civilians, it was really only a matter of time and circumstances before the army would sanction the murder of its own troops. The extent to which the officers and soldiers of the Wehrmacht were willing to carry out such orders and collaborate in the killing of their comrades without any legal proceedings was merely one more measure of their brutalization, a process begun with Hitler's "seizure of power" and reaching its climax on the Eastern Front.

Harsh discipline thus played a far more important role in preserving unit cohesion than "primary groups." It was particularly effective within the context of a brutal war in which soldiers were not only ordered to commit crimes against the enemy, but also allowed to get away with breaches of discipline toward prisoners and civilians. Indeed, the preservation of cohesion and the prevention of rebellion made such unpunished offenses a "necessary" aspect of the war. The "legalization" of authorized, and the toleration of unauthorized crimes was a central component of the Wehrmacht's remarkable determination on the battlefield, quite apart from the sense common to so many soldiers and officers of having to stick together precisely because they shared a common guilt. This mechanism could function properly only because of the general sympathy felt by all ranks for the major tenets of Nazi ideology which made Soviet prisoners and civilians into such ideal targets. Yet although this was an effective means to preserve cohesion, it simultaneously secured the Wehrmacht's final defeat, for as a growing number of officers came to realize, the Soviet regime could only be toppled with the collaboration of the Soviet peoples, a condition made impossible by Germany's murderous occupation policies.

All this is not to say that the *Ostheer* was entirely immune to breakdown. Only four days after "Barbarossa" was launched, the commander of XLVII Panzer Corps expressed concern about the

> repeated *occurrences of panic,* particularly among the baggage and supply column, upon hearing reports (which have proved to be false) of approaching enemy tanks. Whole battalions have turned around on their tracks and fled back to the rear. *This behavior is unworthy of German soldiers.* I charge the unit commanders with the responsibility for preventing any repetition of these occurrences of panic.... It is especially unworthy, however, that it was the officers themselves who had given the signal for this type of oc-currences of panic. In the future such officers are to be immediately reported or arrested by the commander or superior officer in the vicinity on the grounds of cowardice in the face of the enemy.... If it is found that German soldiers have been spreading panic, they are to be immediately brought before a court martial.[171]

The commander of the 18th Panzer, which as part of this corps experienced several such incidents, used a similar tone in addressing his officers:

> Every formation, unit, or column commander, who violates the repeated orders concerning discipline, should be clear in his mind that he is seriously endangering the course of the fighting and is thereby taking upon himself a heavy responsibility.[172]

But while the initial successes of the campaign meant that these instances of panic presented no threat of disintegration, during the latter part of the war they became cause for much greater anxiety among commanders. The 18th Panzer, for instance, experienced a long series of breakdowns during the last months of its existence in summer and autumn 1943. On the first day of the *"Zitadelle"* of-fensive one of its rifle regiments refused to assault the enemy's positions when it came under a heavy artillery barrage, and a re-connaissance company which retreated in chaos from an abortive attack caused a general panic among the men of another rifle bat-talion which culminated in a hasty withdrawal from a well-defended line.[173] Not long thereafter the *Alarmeinheiten* raised by the division as an emergency measure fled from the battlefield the moment their officers were hit. The next day one of the rifle regiments "came out of control" in the wake of a Soviet attack, the troops fled to the rear, and "within minutes the divisional headquarters itself became the front-line." A few days later another infantry battalion fled as a Soviet tank attack rolled over its positions.[174] The divisional commander was highly distressed by this series of incidents whereby, in his own words, "companies, on hearing the cry 'enemy tanks,' spring on the vehicles and tow-tractors of the heavy weapons and drive away to the rear in wild confusion." We have already seen that unlike the manner in which such cases were treated in 1941, by now officers, NCOs, and even privates were expected to shoot down without further ado anyone suspected of causing panic.[175] But the divisional commander also tried to harp on his men's professional and national pride, urging them:

> Preserve your feeling of unconditional superiority over the Russian infantry, who have always been inferior to you and remain so now. Fight hard and resolutely against any manifestation of panic! Do not let yourselves be induced to abandon your anti-tank fox-holes and keep in mind how little the enemy can see from his tanks. Maintain an iron discipline among your ranks, carry out your duty even when your commanders have fallen. He whose heart is in the right place, can and should lead even without shoulder-straps and stripes.[176]

Ultimately all this was to no avail. The 18th Panzer was in the midst of a particularly acute manpower and material crisis, and its disbandment in October 1943 may well have had to do with these frequent cases of disintegration. Yet the *Ostheer* as a whole maintained its cohesion for almost two more years, fighting with a great deal of devotion and determination. Only when everything was lost did the soldiers finally give up, and even then their officers still tried to keep them in their positions. The GD Division, for one, disintegrated only in April 1945. Coming under attack by a far superior Soviet force, Captain Mackert, a GD battalion commander, saw a nearby unit climb out of its positions and march to the rear. Soon one of his own companies, first on the Soviet line of attack, also fled from its positions. "All my attempts to keep the company together failed," wrote Mackert, who was threatening them with his pistol: "The men would rather be shot than stay in their positions." Within a few moments the captain was left with only a company sergeant, two radio operators, and a runner. He never saw his men again.[177]

From all this it is clear that at critical moments, when terror from the enemy became even greater than fear of one's superiors, incidents of breakdown among combat units did occur, and no amount of disciplinary brutality could prevent them. But what is most important about these incidents is that although they were far from infrequent, at no stage of the war save for the very last weeks did they threaten the cohesion of the army as a whole. Thus it was shown that just as brutal discipline could be accepted by the troops only because they had been taught to believe in the ideological arguments on which it was based, so too this ideological cohesion of the troops assumed a major role in preventing the organizational disintegration of the army when the disciplinary system crumbled. Paradoxically, while discipline was aimed at instilling into the troops fear of their superiors, indoctrination increasingly terrorized the soldiers by horror tales about what they could expect from the "Judeo-Bolshevik" and "Asiatic flood" threatening the cradle of culture. Thus precisely when fear of the enemy in one point of the front overcame fear of punishment and caused local breakdowns, the overwhelming terror from the ultimate consequences of a Soviet victory rapidly isolated this incident; rather than the breakdown spreading across the front, the reaction of nearby units was to steel themselves once more and make yet another effort to halt the demonic hordes advancing from the East. Mutiny and disintegration tend to have a contagious effect on armies and to spread with remarkable speed; the Wehrmacht protected itself from most breakouts by harsh discipline, but completely inoculated its troops from a panic epidemic by huge counter-injections of terror from the enemy. Indeed, one can say that the typical *Landser* was a very frightened man, scared of his commanders, terrified of the enemy; this is probably why he seems to have enjoyed so much watching others suffer. The photographs of smiling Wehrmacht troops, each with his little camera, busily taking pictures of hanged "partisans," or of piles of butchered Jews, this horrific *"Exekutions-Tourismus,"*[178] can only be understood as the ultimate perversion of the soldiers by a terroristic system of discipline, backed by a murderous ideology, which achieved its aim of preserving cohesion at the price of destroying the individual's moral fabric and thereby making possible the extermination of countless defenseless people. The troops' perception of reality and understanding of their actions was distorted by the conditions and circumstances of their existence. Yet it must be emphasized that it was the years of premilitary and army indoctrination which molded the soldiers' state of mind, prepared them for the horrors of war, and instilled into them such determination and ruthlessness. The following chapter will examine the extent to which Nazi ideology shaped the Wehrmacht into Hitler's army.

4

The Distortion of Reality

In mid-July 1941 a Wehrmacht NCO wrote home from the Eastern Front:

> The German people owes a great debt to our Führer, for had these beasts, who are our enemies here, come to Germany, such murders would have taken place that the world has never seen before. . . . What we have seen, no newspaper can describe. It borders on the unbelievable, even the Middle Ages do not compare with what has occurred here. And when one reads the *"Stürmer"* and looks at the pictures, that is only a weak illustration of what we see here and the crimes committed here by the Jews. Believe me, even the most sensational newspaper reports are only a fraction of what is happening here.[1]

This striking inversion of reality, which ascribed the unprecedented brutality of the Wehrmacht and the SS to their victims, was the most characteristic feature of the German soldier's "coming to terms" with his actions in the Soviet Union. Indeed, it can be said that this was probably the most effective means of overcoming the moral scruples many of the Wehrmacht's troops and officers may still have retained in spite of their long years of ideological training. Yet it is precisely this distorted perception of reality which gives us the measure of success of Nazi propaganda and indoctrination.[2] The German soldier was not expected to become ideologically committed to some sort of National Socialist theory or dogma; although the regime naturally wished the soldiers to accept its *Weltanschauung,* this never claimed to be a consistent and coherent set of ideas, but rather was a series of slogans meant to derive from and guide one's actions. The central themes of Nazi ideology, racism and especially anti-Semitism, anti-capitalism but much more so anti-Bolshevism, expansionism and most particularly the creation of *Lebensraum* in the East, as well as the construction of a harmonious and "racially pure" *Volksgemeinschaft* at home, ruled, however, in accordance with the *Führerprinzip,* had much affinity with the army's own modes of thinking, principles of organization, self-appointed tasks, and ideals.[3] Both Nazism and the military tended to idealize battle as the supreme test of the individual, and to view the comradeship in arms of the soldiers, the so-called *Kampfgemeinschaft,* as the perfect model for social organization. This was no coincidence, for just as many of the Nazi ideas originated in the military, so it seemed to a growing number of officers only natural to reintroduce them back into the army in their more radical, Nazified form. National Socialism's tendency to stress action, rather than theory, also nicely corresponded to the military mentality. The deed proceeded the thought, constantly molding and confirming it.[4] Yet this was not a pragmatic approach, for underlying the action there were a few essential and unchangeable beliefs and dogmas which no amount of empirical evidence could disprove, particularly as they were not based on rational examination or logical construction, but on faith. Hence, in order not to be hampered either by disturbing manifestations of reality, or by previously acquired moral codes and traditions, one had constantly to observe the world, and particularly the causes and consequences of one's actions in it, through Nazi ideological filters. This tended to become a self-fulfilling prophecy, whereby acting in a manner perceived as necessary for the situation one expected actually created that situation, confirming one's expectations and justifying one's actions. The confusion between cause and effect was indeed inherent to the war in the East, and constituted a vital component of its destructive energy. It also implanted itself deeply in the German collective memory of the war. The distorted features of the tortured and butchered served as evidence of their own, rather than of their murderers' inhumanity; the sense of moral outrage and physical disgust they aroused produced a powerful desire for revenge, which by a process of inversion was directed at the victims rather than the perpetrators, that is, the "other" rather than oneself, for it was their presence which had made such atrocities necessary, their evident inhumanity which had revealed one's own barbarity. Hence, only by physically annihilating the victims and erasing their memory could one salvage one's own humanity.[5]

Such a profound distortion of one's rational faculties and perception of reality, such an effective mechanism of inversion, called for a major indoctrinational effort. To be sure, soldiers often cope with the destruction of enemy lives and property by shifting the responsibility for their actions to their opponents. Dehumanizing the enemy is an inherent element of war; when soldiers attach a particular identity to the individual they confront on the battlefield, pulling the trigger may become all but unbearable. Only in rare cases do soldiers kill out of personal hatred for a specific individual; rather, they may sense hatred or wish to take revenge on a generalized, faceless entity who constitutes the "enemy." Most often, soldiers find it easier to kill each other precisely because they do not perceive the enemy as a fellow human being. Indeed, soldiers confronted with the realization that they had actually killed individuals not unlike themselves can be struck by a sense of personal guilt accompanied by outrage at those who had ordered them to do the killing. Nonetheless, the case of the Wehrmacht, and especially of the *Ostheer*, was essentially different, for this was no ordinary war between two opposing armies, but a campaign of murder and destruction which dispensed with all previously accepted norms of conduct, and intentionally mobilized the unavoidable sense of guilt for killing innocent civilians and unarmed soldiers as an engine even further to enhance its barbarity, punishing its victims for having made their persecutors into monsters. Yet although the conditions at the front and the nature of the occupation accelerated this process, its roots lay in a much longer preparation going back to the soldiers' prewar, premilitary experiences.

Most of the men who served as the Wehrmacht's combat troops during the Second World War were either children or teenagers when Hitler came to power in 1933. An eighteen-year-old soldier in 1943 would have been only eight during the Nazi "seizure of power." Thus the fighting spearhead of the Third Reich was composed of men who had spent the formative years of their youth under National Socialism. This was of particular importance because the regime was first and foremost concerned with indoctrinating Germany's young generation, both in the official educational system and especially within the ranks of the *Hitlerjugend* and the *Arbeitsdienst*. Exposed to the influence of these new and still attractive institutions at a highly impressionable age, there is little doubt that the youths who were to become the Wehrmacht's combat troops were to a large extent molded in the spirit of Nazism, and prepared for the kind of war the regime was determined to wage. The *HJ* gained much of its appeal by openly opposing the traditional foci of authority, the family and the school, and by presenting itself both as a rebel youth movement set upon destroying a staid and anachronistic present so as to create a bright future full of adventure and promise, and as a highly disciplined and devoted body of followers united by a single cause and led by a quasi-divine leader. This combination of rebellion against the old forms and institutions, and an exhilarating, fanatic devotion, this commitment to destroying the present and building the future, this desire to disobey the old rules and rulers and yet "blindly" to follow the new, to act rather than think, this celebration of youthful vitality accompanied by a fascination with death, was at the very heart of the Nazi takeover of Germany's youth and future soldiers.[6]

The manner in which National Socialist indoctrination molded the minds and personalities of German youngsters is vividly portrayed in their memoirs. In the case of Alfons Heck, who was six years old in 1933, membership in the Hitler Youth was so important that it constitutes the main theme of his autobiography. Raised by Catholic parents without any Nazi inclinations, Heck soon joins the *HJ* and becomes a fanatic supporter of the Führer. Significantly, although the events described in the book are set in the same rural region of the Rhineland as the film saga *Heimat*, while the protagonists of the latter are supposed to have remained almost completely untouched by Nazism, Heck's is a *Bildungsroman* concerned with the making of an innocent child into a zealous Hitler Youth leader whose greatest desire is to sacrifice himself for the Führer.[7] Naturally, even in 1944 the seventeen-year-old Heck knows little National Socialist theory; but as a fighter for the cause he embodies the *Idealtyp* of the regime. Although Hitler's suicide sweeps away all his previous loyalties, as long as the Führer is alive the lad's devotion to him constitutes the most crucial motivating force of his young spirit. One may only wonder what traces this youthful, twelve-year long addiction had left on the adult's postwar character.[8] The slightly younger Dieter Borkowski, raised by his mother in the great metropolis of Berlin, also describes his identity as having been molded mainly by the *HJ*, anti-Semitic films, and the popular *Wochenschauen*.[9] In May 1945, when this sixteen-year-old lad is still fighting in Berlin with all the determination of a sworn defender of the regime, he hears that the Führer has taken his own life. The tremendous shock he experiences upon receiving this news gives us a measure of Borkowski's devotion to Hitler and everything he embodied for him, however limited his understanding must have been of National Socialism's ideological and practical implications:

> These words make me feel sick, as if I would have to vomit. I think that my life has no sense any more. What was this battle for, what were the deaths of so many people for? Life has apparently become worthless, for if Hitler has shot himself, the Russians will have finally won. . . . Has the Führer not betrayed his *Volk* then after all?[10]

There were many more such cases. The regime won the loyalty of Germany's children and youths by entrusting them with tremendous destructive powers. Some were soon old enough to exercise these powers against the Reich's real or alleged exterior enemies; the younger generation was confined to the family circle and the school, where it functioned as the regime's corps of agents and informers. Thus the twelve-year-old Jochen Ziem, and the sixteen-year-old Karl Hillenbrand, who spent an idyllic childhood in a Siegerland village, came close to denouncing their parents to the Nazi authorities when they either clashed with them on family matters or felt they were betraying the Reich.[11]

It was precisely the young combat soldiers of the Wehrmacht who were most likely to have come under the influence of Nazi indoctrination before their conscription. Children aged between six and eleven in 1933 spent long years in the *HJ*, yet had a good chance of finding themselves at the front during the war. Interestingly, social backgrounds seem to have played a far less important role in deciding these youths' position toward the regime than we would expect from patterns of support to the Nazi party before the "seizure of power." Eugen Oker, who grew up in a Catholic Bavarian village, was fourteen when Hitler came to power, and that very same year became an ardent *HJ* follower.[12] Gustav Köppke, who grew up in the Ruhr industrial region, and both whose father and stepfather had been miners and communists before 1933, likewise became a devoted *HJ* member. He was only nine when he watched the *Kristallnacht* pogrom: "It was terribly impressive, when the SA marched. . . . I was on the side of the strong guys; the Jews, they were the others." For boys of his age, Köppke reported, being of working-class origins mattered very little:

> Our workers' suburb and the *HJ* were in no way contradictory. . . . this idea of the *HJ* versus the people, you shouldn't see it as if we young lads had to decide for something or against something; there was nothing else. . . and whoever wanted to become something belonged to it. . . . The *HJ* uniform was something positive in our childhood.

Strongly influenced by wartime propaganda, Köppke considered partisans as *Untermenschen,* and domestic criticism of the regime as treason, coming close to denouncing his parents for speaking with Polish *Zwangsarbeiter.* In 1944, when he was hardly sixteen, Köppke volunteered to the SS *HJ*-Division, and was just as shocked as many of his contemporaries by the collapse of the "thousand-year Reich" not long thereafter:

> I was raised then, in the National Socialist time and had seen the world just as they had shown it to us. . . . And suddenly nothing made sense any more.[13]

Another son of a working class family, Gisberg Pohl, already held a senior *HJ* position when he volunteered for the *Waffen*-SS at the age of eighteen in 1943. He did his basic training in the concentration camp Buchenwald. Forty years later he recalled the effect of what he saw there: "For me a whole world came apart," he maintained, especially because "I was then . . . quite earnest," and "[a]lthough they naturally tried to explain to us . . . that these were *Untermenschen,* Russian POWs, Jews, I don't know who they rounded up there." Yet from the perspective of late middle-age Pohl maintained that "I naturally made too much of it then, right, and one has made too much of it later." This son of the working class also took part in the suppression of the Warsaw rising, and remembered asking himself at the time: "Does the Führer know this then?" But once more he hastened to add to his interviewer that

> being a young man one easily made too much of it. We had after all gone to Russia, we wanted there [to destroy] subhumanity—I was, that is, strongly convinced of my task, that I was right. And once it goes that far, then you don't think about it much, then only one thing remains, then you know very well, either him or me.[14]

There were of course exceptions. As we learn from Heinrich Böll's autobiography, certain families did retain their children's loyalties and succeeded in preventing their becoming enthralled by the *HJ*.[15] Yet even in Böll's case it should be remembered that he served for six years in the Wehrmacht, mostly on the Eastern Front, and though wounded on numerous occasions, kept returning to the fighting. Moreover, as some of his early postwar stories reveal, he too was not entirely free from the stereotypical views of the enemy pumped into every German youth and soldier, combining for instance the propagandistic image of the insidious/partisan/whore, with his own brand of the innocent/angelic/virgin which featured later so prominently in his mature writing.[16] Bernt Engelmann, though conscripted to the Wehrmacht, collaborated with the resistance and ended up in a concentration camp. Yet in his book on that period he writes:

> It was not uncommon for young people in Germany to reject their families' convictions and follow the Nazis, and many fathers and mothers, themselves strongly opposed to the Hitler regime, were denounced by their own children as enemies of the state or even turned over to the Gestapo.[17]

Indeed, as he travelled around Germany forty years after the event trying to reconstruct people's memories of the Nazi dictatorship, he found out that many of his contemporaries could still think of those years with a great deal of nostalgia. One of his former schoolmates, for instance, had this to say on the first wave of terror in March 1933:

> I just remember how thrilling it was. People were electrified, and they all talked about the "unity of the German people," and the "national uprising." I loved the constant marching and singing,

with flags and bunting everywhere. . . . All in all we had a wonderful, carefree youth, didn't we?[18]

His friend also managed to marry a Wehrmacht officer on none other but 9 November 1938, but asked whether she had not been upset by the events of *"Kristallnacht"* that very day, she said: "When we left the banquet that night, the streets were littered with broken glass. And of course I had on delicate evening slippers, and a floor length gown. Father said it was a disgrace that the crews hadn't cleared up the streets yet."[19] But while this woman simply went on leading a normal life, hardly noticing the changes taking place around her as long as they did not touch her personally, others were far from indifferent, especially the young men who ended up in field-gray uniforms. In 1981 Engelemann overheard a man in his sixties reminiscing about his youth in a manner not unfamiliar to anyone who has frequented German *Kneipen*: "Back then we hadn't the slightest doubt about the rightness of our cause or the certainty of the ultimate victory. That's what we grew up with—at home, in school, and in the Hitler Youth." It later turned out that he had been a tank driver in the Adolf Hitler *Waffen*-SS Division.[20]

Hannsferdinand Döbler, who as a twenty-six-year-old in 1945 was of the same age group as Böll, was far more typical of his generation than the celebrated author. Describing himself in his memoirs as a "150 percent idealistic-believing officer," he kept fighting even after the capitulation was formally announced. Significantly, although he totally internalized the regime's value system and conformed to the new ideal type of the Wehrmacht's combat officer, Döbler did not consider himself a party member; indeed, just like the younger Heck, he saw himself as being of a higher quality than the staid and corrupt *"Alte Kämpfer."* His energetic and self-sacrificial devotion to the regime was expressed in a will to conform to its models of heroism and action. Raised by his mother in a petit-bourgeois family, Döbler wished more than anything else "to belong" and "to be there." Unlike Böll, his pastor seemed to him pathetic, and his friendship with a half-Jewish girl had no impact on his anti-Semitic views. Döbler's identity was molded in a constantly military environment where action replaced thought. He was indeed an outstanding example of the type manufactured by that powerful combination of the Nazi regime's ideology, the Wehrmacht's system of values, and the reality of the war, enhanced by the youthfulness of the soldiers, the manifest weakness of family and school in the face of totalitarian rule, and the tremendous impact of a highly appealing youth movement, which deliberately mobilized the rebellious spirits of the young against their parents and teachers, providing them instead with military trappings, power over their elders, and an opportunity to sacrifice themselves for a "good cause."[21] Such idealistic junior officers in turn had a substantial influence on the rank and file, while simultaneously generating a feeling of pride in the Wehrmacht's achievements among the population in the rear. A good example of this effect can be seen in a letter written in August 1941 by Frau Else Gaupp of Bad Cannstatt to the commander of the 18th Panzer Division, then in the midst of the Russian campaign. Thanking the general for the kind words with which he had notified her of her son's death in action, she added: "You, Herr General, knew indeed only the skilful officer Julius Kirn, but he was much more than that: he had character, a beautiful, harmonious man, a fully chivalrous son." So as to demonstrate her son's devotion to the cause, she then went on to cite one of his last letters, in which he spoke of his feeling

> of being able to step forth in battle as a leader, to return from a hard, but victorious fight, to see and to experience how the eyes of the soldiers are directed at me, while the artillery shells are exploding all around, and to know that one can depend on oneself, because in spite of the anxiety one is capable of doing and achieving the right thing, [and] this knowledge and this experience make me proud, and I am endlessly thankful for my fate, which has allowed me to experience such hours. Some scrutinize their anxiety, I too have fear, [but] it is beautiful to overcome it. Perhaps precisely we Germans possess the ability to be happy in battle, because it is especially the fighting which provides us soldiers with the best measuring-rod, and each of us comes to know himself, if he is truthful with himself. If I fall, this death will also be an experience for me, it will be the last fulfilment of my profession and of my life.[22]

Similarly, in 1987 Mrs. Helene Fuchs Richardson introduced a collection of letters written by her husband, a Panzer sergeant, platoon commander, and officer candidate, killed near Moscow in November 1941, with the following remarks:

I assume it was his artistic nature and his enthusiasm for the new National Socialist Germany which first brought him to the Hitler Youth and then into the Nazi Party.... Karl was very much in favor of the National Labor Service.... When he was drafted for military servive, he volunteered for the tank corps because this unit seemed to be most important to him. His letters from the front speak of his deep love for the Fatherland and the strong belief in the Führer.

He was born in 1917 and as a youth grew up with the teachings of National Socialism. He was an impressionable young man and, like many of his contemporaries, was overwhelmed by the preaching of the party.... Since he was such an idealist and a firm believer in the Third Reich, it would have been a bitter disappointment for him to have experienced the total collapse of Nazi Germany in 1945.[23]

And his son, whom he had never seen, adds: "My father's letters from 1937 to 1941 provide a unique perspective of this important period from the point of view of a typical, young German man. He was truly devoted to the cause of Greater Germany and felt it was his sacred duty to engage in battle for this cause."[24] Indeed, Karl Fuch's devotion to the Nazi cause was manifested early on, long before he found himself on the battlefield. On 23 November 1938, while still a student, he wrote his parents:

By God, you should have been in Würzburg during this Jewish mess [*Kristallnacht*]. I don't know if things were as hectic in Nuremberg, but we made a clean sweep here. I can tell you that the authorities didn't miss one of those pig Jews.[25]

There were many such enthusiatic young men among the Wehrmacht's soldiers and junior officers, powerfully imbued both with racist sentiments and with the notion that war was the climax of human existence. Karl Fuchs wrote his wife in May 1941: "Life, by definition, means struggle and he who avoids this struggle or fears it is a despicable coward and does not deserve to live."[26] A few days later he added: "The struggle for existence... creates proud, free, honest and upright people. All others will remain repulsive creeps, inferior individuals who shy away from danger and who, when the chips are down, will succumb to that danger."[27] Heinz Küchler, a twenty-six-year-old student of law, wrote from the Eastern Front on 11 July 1941:

Sometimes conversations in a small group revolve around [contemporary] events and the future; no outlook seems promising; only fresh individual courage for life will know how to master future times, in which lying ideals, false gods and untruthful wisdom will be and must be smashed. Calm and peace, tranquil happiness will no longer be alloted to this generation; war will go on for many years, if not for centuries! Oh well, one will not be put to sleep and will not rust, will not find boredom, will not be satisfied with phrases and flattering lies; perhaps precisely this time will lead toward truth and knowledge.[28]

And on 6 September he added:

I understand the endeavours of men to give their death a meaning, in that they conjure for themselves a picture of the soul, of a battle for a great, just, holy cause. . . . My own battle is different. . . . The struggle for the truly human, personal values, for the timeless cause of the spirit, the spirit that creates the tie between man and God, in constant endeavour for knowledge and truth. And this battle must be fought today also materially, with sacrifices of blood and life.[29]

Eberhard Wendebourg, a student at a teachers' seminar aged twenty-four, wrote from the East on 5 October 1941: "For us this means now only to go on doing our duty faultlessly, independently of any recognition and tribute. This is a hard, but valuable school, to have to accomplish His work, His task. But how it fortifies the soul, how it steels the will!" Wendebourg looked forward to returning from the war and teaching what he had learned there to his students: "It will be wonderful to work then. . . . And the *Volksgemeinschaft*, a truthful goodness and love among all Germans, will be fought for anew, even better than in the years before the war." His greatest hope was "to show German youths German values and German greatness and to educate them into real Germans, in whom spirit and mind, will and soul are equally well learned."[30] Günter von Scheven, student at the Arts Academy of Berlin aged thirty-three, wrote on 18 August 1941 from Russia: "What makes me strong is the realization that each sacrifice is necessary, because it is tied to the necessity of the whole."[31] He then added in December: "This [war] is about overcoming the chaotic passage and the preservation of human dignity, which is purified by pain and renunciation. . . . We are fighting not for political contestables, but in the belief that the Noble and the Best must prove itself anew in the battle with the ghostly manifestations of Materialism. I see the whole nation in the process of being recast, in a storm of suffering and blood, which will enable us to reach new heights."[32] The Alsatian volunteer Guy Sajer describes in his memoirs how his company commander's "obvious and passionate sincerity affected even the most hesitant." This officer's moral code made for great devotion to his own men at the expense of everyone else: "I would burn and destroy entire villages if by so doing I could prevent even one of us from dying of hunger"; his ideals were so great that they could only be accomplished by a sort of universal plastic surgery: "We are trying . . . to change the world, hoping to revive the ancient virtues buried under the layers of filth bequeathed to us by our forebears"; this operation must be brutal, and if it fails, "those of us still alive . . . will be judged without mercy . . . accused of an infinity of murder . . . spared nothing"; consequently, the soldiers should remember that "life is war, and war is life. Liberty doesn't exist." Upon hearing this nihilistic/idealistic speech by their company commander, Sajer tells us, "we loved him and felt we had a true leader, as well as a friend on whom we could count."[33] Yet these words could have just as well been spoken by Hitler. The Führer and the junior officers indeed spoke the same language, sharing a paternalistic devotion to the *Volk* or "their men," a commitment to use the most ruthless means against anyone perceived as an obstacle to the realization of their vague, distant, but to them highly idealistic goals, and deriving the same sense of acute joy from the actual act of destruction, from the killing of others and the prospect of one's own death.

The premilitary preparation of Germany's youth combined organizational principles such as group loyalty and absolute obedience to superiors, individual qualifications which put physical stamina far above intellectual ability, and certain ideological tenets, the most important of which were racism, anti-Bolshevism, expansionism, and a quasi-religious faith in the Führer, who constituted both the supreme authority, the embodiment of the *Volk*, and the arbiter of Germany's destiny. As the youth movement adopted military forms of organization, discipline, and training, on the one hand, and the army introduced ideological indoctrination to the troops, on the other, conscription was not experienced as a move to a fundamentally different environment. As far as Germany's military tradition went, however, the Wehrmacht chose a radically new path by its decision to devote a considerable effort to the political, that is, National Socialist education of the troops, and its particular insistence on instilling into them a mystical belief in Hitler. Indeed, it is indicative that only in the Army and the *HJ* every individual swore a personal oath of allegiance to the Führer;[34] and, excluding the SS, these two organizations did in fact stand out as more devoted to Hitler than any other group in the Reich. The indoctrination of the soldiers was of crucial importance in two related ways. First, it taught the troops totally to trust Hitler's political and military wisdom, and never to doubt either the morality of his orders or the outcome of his prophecies. Indeed, the attachment to the Führer it created was so powerful that only his death could finally break it. Second, it provided the soldiers with an image of the enemy which so profoundly distorted their perception that once confronted with reality they invariably experienced it as a confirmation of what they had come to expect. Indoctrination thus served the double purpose of strongly motivating the troops and greatly brutalizing them, for it legitimized both one's own sacrifices and the atrocities committed against the enemy. Faith in the Führer allowed one to believe in the essential moral value of the most heinous crimes, and to trust his promises of the inevitable *Endsieg*; a false perception of reality made for viewing the enemy as a host of political and biological demons, and to ignore all signs pointing at the approaching military catastrophe.[35]

"Belief" in Hitler, in an increasingly religious, metaphysical sense of the term, was a central element in Nazi ideology. Its achievement was facilitated not only by Hitler's own tendency to refer to himself as the representative of providence in the world,[36] but also by a general susceptibility to such appeals by large sectors of German society. As one historian has written:

> Especially in Protestant Germany, Hitler's claim to be a providential saviour had a strong resonance. His religious invocations suited a society which for generations had seen the interwining of the divine and the secular. As religious faith waned, the secular (the nation or the *Volk*) was sanctified. Above all one was taught to sacrifice for the state, as once one had sacrificed for God, and yearn for the community of the *Volk*, as one had once been given strength by a congregation of the faithful.[37]

It is indeed striking to note that as early as October 1932, even before his fatal decision to collaborate with the NSDAP, the otherwise not particularly perceptive von Papen realized the pseudo-religious, dogmatic, and totalitarian essence of Nazism. To his mind, what gave

> the doctrine embraced by the NSDAP . . . the nature of a political religion [*Konfession*] is its axiom of the "exclusiveness" of the political "all or nothing" [and] its mystical Messiah-faith in the "word-mighty" [*wortgewaltigen*] Führer as the only one summoned to control destiny. And indeed it is here that I see the unreconcilable difference between Conservative Politics rooted in Faith and a National Socialist Faith rooted in Politics.[38]

There are many examples of the power of this faith in Hitler, in some cases lasting well after the collapse of his regime. Bernt Engelmann cites a woman of his generation who described her mother as one of

> those who believed in the Führer as the saviour and were hypnotized by him. . . . She was convinced everything the Nazis did was right and essential, and she dismissed all whispered rumours of atrocities as stupid, malicious gossip. . . . In May 1945 her whole world collapsed. . . . Mother was among those the Americans forced to tour nearby Dachau. . . . Mother suffered a nervous breakdown. . . . [But] nothing could shake her faith in Hitler. "I'm sure the Führer wouldn't have wanted *that*," she said later. . . . "True National Socialism was pure and decent!" She clung to that till she died, only three years ago.[39]

And another of his acquaintances, Frau Gussi Hohlbaum, similarly claimed in the early 1980s to have remained true to the flag "to the bitter end," and insisted that though

> mistakes had been made . . . to this day I'm absolutely sure the Führer himself never wanted these things and probably didn't know about them. . . . And yet he really did accomplish the impossible! Millions of people found new happiness.[40]

Yet the purpose of the Nazi sacralization of ideology was not merely to achieve abstract belief, but rather to harness faith as a motivating engine for concrete action. This was stressed time and again by the regime's ideologues, who frequently associated the term *politischer*

Glaube both with a "blind" subordination to the Führer, and with the "deed." In 1938, for instance, the SS journal *Das Schwarze Korps* asserted in an article entitled *Kult und Glaube*:

> True and vigorous faith cannot exist in the abstract, it reaches its fulfilment only in the deed. The deed is the only true witness to faith . . . faith-movement-action . . . these are the three terms that determine for us the natural path of human piety.[41]

In fact, as early as 1927 Hitler himself pointed out that he had no interest in understanding, but only in belief, for it was thus that one could achieve total commitment:

> Be assured, we too put *Glauben* [faith] in the first place and not cognition. One has to be able to believe in a cause. Only *Glauben* creates a state. What motivates people to go to battle and to fight and to die for religious ideas? Not cognition, but blind faith.[42]

The Wehrmacht was well aware of the powerful need for belief among soldiers living in conditions of constant danger, and catered to it with an endless stream of leaflets, brochures, speeches, radio talks, newspaper articles, and all other forms of propaganda directed at the troops throughout the war. It is of some interest to cite a few examples from this material, for it is only by reading these texts that one may grasp the extent to which they relied on emulating religious formulas, on the one hand, and served the troops as a model for articulating their own feelings in private correspondence, on the other. The Wehrmacht's propaganda made a conscious and concerted effort to associate Hitler with God, to present "his mission" as emanating from a divine will, and to tie his personal fate with that of the German *Volk*, indeed with the destiny of "Western civilization" as a whole. Political faith both motivated people into concrete action, and was simultaneously enhanced by it. Note, for instance, the excerpts from a speech made by Hitler in April 1940 which the army distributed to the units as posters to be hung in the barracks. The unmistakable eschatological tone of this speech, which combined pseudo-theological arguments with social-darwinism,[43] and tied together military traditions and Nazi ideology, was doubtlessly intended to create an impression of divine intervention. Indeed, as we shall see below, this unholy mixture did have a powerful effect on the troops. The German soldier, Hitler asserted, is the first representative of life in this struggle [for existence], for he has always been the best selection of those people who by their life's mission and—when necessary—by their loss of life have ensured the life of the others in this and thereby in the next world. . . . No one who has not added to tradition by his own life and action may speak of tradition. . . . Whatever is to be with the life and destiny of the individual, supreme to all of them is the existence and future of the whole . . . for us all has been revealed what so many will still certainly have to fight for in the near future: the German *Volk*! The world wishes our dissolution. Our answer can only be the renewed oath to the greatest community of all times. Their aim is German disunity. Our creed [*Glaubensbekenntnis*]— German unity. Their hope is the success of capitalist interests, and our will is THE NATIONAL SOCIALIST VOLKSGEMEINSCHAFT![44]

Similar terms were used in the radio program "The Voice of the Soldiers," regularly delivered "by an officer of the Wehrmacht." The twelfth lecture, broadcast on 25 April 1940 and issued simultaneously in printed form, was entitled "Arms, Comradeship and Total Commitment [*Einsatz*]." Filled with enthusiasm at the exploits of Germany's armed forces in the Scandinavian expedition, the speaker asserted that

A new spirit distinguishes the present war. Every soldier, every military leader, indeed everyone in Germany has the same goal in mind and is straining to do all in his power in order to fulfil it. We are experiencing what is called "total war-leadership". . . . This is based first and foremost on the fact that everything is guided and brought into unison by the will of *one personality*.

This will, the presence of the strong personality of the *Führer*, is of decisive importance. Providence [*die Vorsehung*] has sent Germany the Führer at the right hour. At this moment we view his work over the preceding years from a new perspective: it has created the uniform spirit which has become so strongly visible during the previous weeks in the collaboration between army, navy and *Luftwaffe*. We soldiers know that without the work of the Party this would not have been possible. The Party is the carrier of the spiritual and the mental preparation for the present mighty task of our people.

What does total war-leadership mean? It consists of more than the concentration of the military under a unified supreme command. . . . It can also not be completely explained by the fact that the

Wehrmacht and the economy construct in war a single unity. It demands the same spiritual, mental and ethical orientation of 80 millions. It calls for the same determined will to victory in all positions of public and private life. It owns the unlimited confidence in our cause and the firm belief in Germany, in the German people and in its historical appointment and precondition.

Everyone in the homeland ... on the *Westwall*, on the ships ... in the squadrons flying over England and France, feels that in the entire German people beats only one heart, one spirit of war and will to victory; the German strength is not in the numbers of men, artillery, aircraft or warships. . . . No one has that which National Socialism has created in 20 years.[45]

Thus the technological, organizational, and tactical achievements of the Blitzkrieg were redefined as a spiritual phenomenon, created by the Nazi party and directed by a God-sent Führer. Even in 1940, when Germany's relative military strength was at its height, the soul of National Socialism and the collaboration between Hitler and providence were presented as far more important than professional skill and technological innovation. This celebration of the irrational, this cult of the Führer, reached its peak following the victory over France.[46] An article published in the *Mitteilungen für die Truppe*, a news-sheet issued by OKW and distributed to all Wehrmacht units, clearly reflected the quasi-religious fervor which seized both the army's propagandists and, as we shall see, many of the soldiers on the ground as well:

What the reports of OKW in May 1940 had made known is one single grand poem of German heroism and inspired [*Genialer*] leadership. . . . Any attempt to describe the battles of these three weeks of the Greater German War of Liberation with one word which would equal their greatness, must be admitted to border on the impossible. . . . This battle of annihilation was so great that we can only accept with shocked silence and thankful hearts this act of destiny.

Behind the battle of annihilation of May 1940 stands in lone greatness the name of the Führer.

All that has been accomplished since he has taken the fate of our people into his strong hands!

. . . He gave the people back its unity, smashed the parties and destroyed the hydra of the organizations . . . he decontaminated the body of our people from the Jewish subversion, created a stock-proud, race-conscious *Volk,* which had overcome the racial death of diminishing births and was granted renewed children-prosperity [*Kinderreichtum*] as a carrier of the great future of the Fatherland. He subdued the terrible plight of unemployment and granted to millions of people who had already despaired of the *Volk* a new belief in the *Volksgemeinschaft* and happiness in a new Fatherland. . . .

His genius, in which the whole strength of Germandom is embodied with ancient powers [*mit Urgewalt verkörpert*], has animated the souls of 80,000,000 Germans, has filled them with strength and will, with the storm and stress [*Sturm und Drang*] of a renewed young people; and, himself the first soldier of Germany, he has entered the name of the German soldier into the book of immortality.

All this we were allowed to experience. Our great duty in this year of decision is that we do not accept it as observers, but that we, enchanted [*hingerissen*], and with all the passion of which we are capable, sacrifice [*hingeben*] ourselves to this Führer and strive to be worthy of the historical epoch molded by a heaven-storming will.[47]

As the final victory eluded the Wehrmacht in the endless spaces of Russia, casualties accumulated at a terrifying rate, and the tremendous attrition rapidly eroded German material strength, the tone of the army's propaganda changed from ecstatic to frantic, often verging on the hysterical; technology and skill were now to be increasingly replaced by devotion and fanaticism, rational thought by "blind" belief. Commanders were charged with instilling into the men a new ideological fervor with which to combat the enemy's superior numbers and machines, as well as the Red Army's own apparent ideological motivation. In an appeal made in April 1942 by the officer corps' news-sheet, *Mitteilungen für das Offizierkorps,* commanders were urged to realize that

[i]n the struggle against the capitalism and imperialism of the English and the Americans and against the world-revolutionary theses of the Bolsheviks the weapons of the Wehrmacht alone will not achieve victory . . . [which can be gained] only . . . when the people . . . confronts the political and ideological theses of the enemy with better political concepts. . . . [S]uch an attitude . . . is based on the German people's unshakable sense of loyalty to Führer, *Volk* and Fatherland, the kind of loyalty which remains absolutely firm in the face

of all crises and knows no scepticism. . . . Not only are the economic and power-political bases of our life critically threatened, but the whole spiritual life of the nation, the ethical basis of our cultural and religious concept of the world, truly everything which is great and holy for German men in life and death, all is threatened at the core if we fail to master the enemy. . . . Have the officers burnt this so deeply into their men's hearts, that each of them knows and sees fully and clearly against what devilish game in the world he has been called into action? . . . We know that the Devil has been set loose against our land . . . we are filled with the responsibility to God to defend the land which had been given us, to save His property and to multiply it, and therefore we mobilize not only our weapons . . . but also the weapons of the soul. . . . The military-spiritual [*wehrgeistige*, or ideological] leadership of the soldiers has been added to the officers' duties, because political determination and soldierly feats are a single unity and are indissolubly bound to each other. The more German soldiers are aware of the full extent of the mortal danger which threatens them, the greater will be the conviction and the toughness with which they will confront the dynamics of the Bolshevik revolution with the whole strength of soul and will of National Socialist Germany. . . . In the war, as the Führer has said . . . the nations are being judged in the Godly court of the Almighty. He who survives this trial will be seen as worthy of molding a new life on earth. . . . What a task! . . . The officers of the Führer, and the German soldiers whom they lead, a sworn community of the best men of German blood, carried on by the love, the work and the belief of the German people, are marching to the decision. There beyond hell is burning. May it charge! We shall still win![48]

The Wehrmacht's propaganda thus fed the troops with an ever heavier diet of religious images, portraying Hitler and the Nazi creed as God's instruments charged with protecting German culture and blood, and communism as Satan's servant, unleashed from hell to destroy civilization. Only an unquestioning belief in the Führer and the final victory would save the world from subjugation by the devil's hordes. The German soldier could well imagine what these demonic hordes might wreak on his land, for his own army's actions in Russia had provided him with an appropriate example. Indeed, it was fear of vengeance for the Wehrmacht's barbarities which made this propaganda so effective. But the memory of previous victories still left some hope that the tide would turn, and strengthened the belief that Hitler would once more achieve the impossible as he had done so many times in the past. This too became a constant theme in the Wehrmacht's propaganda, as can be seen from the following excerpt, taken from the booklet "National Political Instruction" handed out to company commanders as a guide for the ideological training of the troops:

> Only the Führer could carry out what had not been achieved for a thousand years. . . . [He has] brought together all the German stock . . . for the struggle for freedom and living space . . . [and] directed all his thoughts and efforts toward the National Socialist education of the *Volk*, the inner cohesion of the state, the armament and offensive capability of the Wehrmacht. . . . When the German Eastern Armies fought an unparalleled battle during the winter of 1941–42 in the snow and ice of the Russian winter, he said: "Any weakling can put up with victories. Only the strong can stand firm in battles of destiny. But heaven gives the ultimate and highest prize only to those who are capable of withstanding battles of destiny." In the difficult winter of 1942–43 the strength of the Führer was demonstrated once more, when . . . he called upon the German *Volk* at the front and in the homeland to stand firm and make the supreme effort. The Führer . . . clearly sees the goal ahead: a strong German Reich as the power of order in Europe and a firm root of the German *Lebensraum*. This goal will be achieved if the whole *Volk* remains loyal to him even in difficult times and as long as we soldiers do our duty.[49]

The Führer was thus presented as the creator of the new German nation and the guardian of its ancient traditions, its source of power and prosperity, of fertility and purity. As long as Hitler lived and the people remained loyal to him, no harm could come to the Reich. But the continued effectiveness of this "Führer cult" in the face of ever growing defeats depended to a large extent on the conviction that confronting the Reich, particularly in the East, was an enemy who combined racial perversity with a demonic ideology. Even more important than providing the troops with an ideal image of their own leadership, *Weltanschauung*, and racial qualities, was the concerted attempt to terrorize them by visions of destruction at the hands of the "Judeo-Bolsheviks," indeed, to convince them that all the atrocities they had committed would be turned against them in case they failed to win. The Wehrmacht's barbarous policies were thus utilized for propagandistic purposes as well, providing a vivid and frightening model of what Germany itself could expect in defeat, especially powerful because for some at least it was perhaps tinged with a sense of guilt, with a feeling that such revenge would be well deserved. As one soldier who had witnessed a massacre of Jews in Lithuania in July 1941 said: "May God preserve us from losing the war, because if revenge [for this] is taken, things will be bad for us."[50] For the troops at the front, however, the embodiment of the enemy's ideological and racial qualities, of his fanaticism and will for revenge, was doubtlessly the Soviet commissar. The *Mitteilungen für die Truppe* painted a particularly nightmarish picture of this monstrous angel of death accepted, as we shall see, by many of the soldiers as an accurate reflection of reality:

> Anyone who has ever looked at the face of a red commissar knows what the Bolsheviks are like. Here there is no need for theoretical expressions. We would insult the animals if we described these mostly Jewish men as beasts. They are the embodiment of the Satanic and insane hatred against the whole of noble humanity. The shape of these commissars reveals to us the rebellion of the *Untermenschen* against noble blood. The masses, whom they have sent to their deaths by making use of all means at their disposal such as ice-cold terror and insane incitement, would have brought an end to all meaningful life, had this eruption not been dammed at the last moment.[51]

Just as the image of the enemy's revenge was rooted in the memory of one's own crimes, so too the destructive qualities ascribed to the commissar reflected National Socialism's own powerful nihilistic urge. Similarly, the consistent refusal to discuss the theoretical content of communism emanated from the Nazi emphasis on action and belief, and the characteristic reluctance to engage in rational discussion; faith, physical coercion, or outright destruction were always preferred to persuasion. This was a language well suited for soldiers. Thus the enemy too was portrayed as a "believer" in the religious sense rather than as a professional or a patriotic defender of his country. And, because the Red Army fought with such determination, it became necessary even further to "fanaticize" one's own troops. This was again highly in line with the inherent tendencies of Nazism, for thriving as it had always done on action, it

could reach higher peaks of intoxication from devastating enemy
territories than from building *Autobahnen* at home; consequently,
the more destructive the war in the East became, the greater energies
Nazism could derive from it. The soldier, whose vocation, however
sublimated, is ultimately to destroy, could only welcome a world-
view which now endowed it with a universal, metaphysical meaning.

The propagandistic image of the enemy seemed to be confirmed
upon actual encounter. It has been said that within Germany the
"disappearance" of real Jews from the cities and towns made it easier
for the population to accept the abstract anti-Semitic portrayal of
"the Jew."[52] On the Eastern Front, however, the troops found no
contradiction between the "Judeo-Bolshevik Asiatic hordes" which
sprang from the propagandists' imagination and the enemy soldiers
they were actually fighting. Nor, for that matter, did encounters
with Jews in the East make them doubt the existence of that mythical
Jew who played such a prominent role in their indoctrination material. Quite on the contrary: reality did not disprove myth, but
was rather molded so as to fit it. Thus one measure of the extent
to which Nazi images permeated the soldiers was their manner to
resolve the contradiction between abstract image and actual appearance by destroying the latter as a manifestation of the former.
The distortion of perceived reality consequently led to a distortion
of objective reality: one saw what one expected to see, and one
smashed it so as not to have to see it any longer. This process was
greatly accelerated due to the fact that such propagandistic portrayals
of the enemy were not disseminated merely by the party's organs
and the army's high command in the rear, but also by combat
commanders at the front, many of whom, moreover, did not view
ideology cynically as one more means to motivate their troops, but
seem to have believed it with precisely that sort of fervor demanded
from the Führer's disciples. Indeed, ever since the collapse of the
Kaiserheer the German officer corps had been searching for a new
set of ideas which would form the crucial link between effective
action and spiritual commitment, both endowing the deed with a
higher meaning and deriving its essence from the experience of combat; now that they had been provided with such an ideology, they
were not likely to take it lightly.[53] Significantly, even officers with
little reason to be enamored with Hitler and his regime often shared
many of the Führer's prejudices, or what they preferred to call his

Feindbild, despite the obvious fact that many of their perceived "enemies" hardly constituted an objective threat to the Reich. Thus, for instance, less than a year after his ejection from the position of commander in chief of the army by means of an outrageous fabrication, and merely a few weeks following the *Kristallnacht* pogrom, Colonel-General von Fritsch wrote in a private letter:

> It is very strange that so many people should regard the future with growing apprehension, in spite of the Führer's indisputable successes in the past. . . . Soon after the [First World W]ar I came to the conclusion that we should have to be victorious in three battles, if Germany was again to be powerful:
> 1. The battle against the working class, Hitler has won this;
> 2. Against the Catholic Church . . . and
> 3. Against the Jews.
> We are in the midst of these battles, and the one against the Jews is the most difficult. I hope everyone realizes the intricacies of this campaign.[54]

Fritsch did lose his faith in the ability of the Nazi regime to "save" Germany earlier than most other generals due to his personal misfortune. Yet his faith in Hitler was merely transformed into fatalism, and while he believed that the Führer would most probably lead the Reich into the abyss, he saw him as Germany's destiny and could thus not even contemplate any opposition to his regime.[55] His more adaptable colleagues manifested greater optimism, and their conviction in Hitler's ideological arguments grew following their triumphs in Poland and the West. Two months before the invasion of the Soviet Union, General-Colonel von Küchler, commander of 18 Army, told his divisional commanders: "A deep ideological and racial abyss separates us from Russia," which was, after all, "an Asiatic state." Therefore, he stressed, "the aim must be to destroy European Russia." Expressing complete agreement with the "criminal orders," von Küchler issued his formation commanders with the following instruction: "The political commissars and GPU-people are criminals. . . . They are summarily to be brought before a field court martial."[56] Colonel-General Hoepner, commander of Panzer Group 4, perceived the coming war in the east precisely in the same terms as the regime's and the Wehrmacht's propaganda. On 2 May 1941 he wrote:

The war against the Soviet Union is an essential component of the German people's struggle for existence. It is the old struggle of the Germans against the Slavs, the defense of European culture against the Muscovite-Asiatic flood, the warding off of Jewish Bolshevism. This struggle must have as its aim the demolition of present Russia and must therefore be conducted with unprecedented severity. Both the planning and the execution of every battle must be dictated by an iron will to bring about a merciless, total annihilation of the enemy. Particularly no mercy should be shown toward the carriers of the present Russian-Bolshevik system.[57]

On the eve of "Barbarossa" the commander of XLVII Panzer Corps urged his troops to keep in mind the historical context in which they were setting out against the Soviet Union:

It is now our task to destroy the Red Army and thereby to eradicate for ever Bolshevism, the deadly enemy of National Socialism. We have never forgotten that it was Bolshevism which had stabbed our army in the back during the [First] World War and which bears the guilt for all the misfortunes our people has suffered after the war. We should always remember that![58]

Once the fighting began, rather than attempt to temper their troops' brutality, many commanders seemed to think that the soldiers were still showing too much compassion for the enemy, and strove to instill into them a greater understanding for, and a firmer will to participate in, the brutalities deemed essential for the victorious outcome of this "war of ideologies." The commander of 6 Army, von Reichenau, thus appealed to his troops on 10 October 1941:

Regarding the conduct of the troops toward the Bolshevik system many unclear ideas still remain.

The essential goal of the campaign against the Jewish-Bolshevik system is the complete destruction of its power instruments and the eradication of the Asiatic influence on the European cultural sphere.

Thereby the troops too have *tasks,* which go beyond the conventional unilateral soldierly tradition [*Soldatentum*]. In the East the soldier is not only a fighter according to the rules of warfare, but also a carrier of an inexorable racial conception [*völkischen Idee*] and the avenger of all the bestialities which have been committed against the Germans and related races.

Therefore the soldier must have *complete* understanding for the necessity of the harsh, but just atonement of Jewish subhumanity. This has the further goal of nipping in the bud rebellions in the rear of the Wehrmacht which, as experience shows, are always plotted by the Jews.[59]

This order served as a model for many other Wehrmacht generals, and was indeed highly praised by Hitler himself, who made a point of distributing it to all combat formations in the East. Borrowing some of Reichenau's phrases, on 20 November 1941 General von Manstein, commander of 11 Army, issued his own, if anything even more radical version:

Since 22 June the German *Volk* is in the midst of a battle for life and death against the Bolshevik system. This battle is conducted against the Soviet army not only in a conventional manner according to the rules of European warfare. . . .

Judaism constitutes the mediator between the enemy in the rear and the still fighting remnants of the Red Army and the Red leadership. It has a stronger hold than in Europe on all key positions of the political leadership and administration, it occupies commerce and trade and further forms cells for all the disturbances and possible rebellions.

The Jewish-Bolshevik system must be eradicated once and for all. Never again may it interfere in our European living space.

The German soldier is therefore not only charged with the task of destroying the power instrument of this system. He marches forth also as a carrier of a racial conception and as an avenger of all the atrocities which have been committed against him and the German people.

The soldier must show understanding for the harsh atonement of Judaism, the spiritual carrier of the Bolshevik terror.[60]

Five days later, the commander of 17 Army, Colonel-General Hoth, provided his own troops with a still more elaborate "analysis" of the historical and ideological context of the war, coming nevertheless to the same conclusion that only by annihilating their racially inferior and morally depraved enemy would they save European and especially German "culture" from Asiatic barbarism:

It has become increasingly clear to us this summer, that here in the East spiritually unbridgeable conceptions are fighting each other:

German sense of honor and race, and a soldierly tradition of many centuries, against an Asiatic mode of thinking and primitive instincts, whipped up by a small number of mostly Jewish intellectuals: fear of the knout, disregard of moral values, levelling down, throwing away of one's worthless life.

More than ever we are filled with the thought of a new era, in which the strength of the German people's racial superiority and achievements entrust it with the leadership of Europe. We clearly recognize our mission to save European culture from the advancing Asiatic barbarism. We now know that we have to fight against an incensed and tough opponent. This battle can only end with the destruction of one or the other; a compromise is out of the question.

Hoth went on to say that "compassion and weakness" as regards the population were out of place, and urged the soldiers to understand the "necessity of the harsh measures against racially foreign [*volks- und artfremde*] elements." One must realize that

Russia is not a European, but an Asiatic state. Each step in this unhappy, enslaved land teaches us this difference. Europe, and especially Germany, must be liberated for ever from this pressure and from the destructive forces of Bolshevism.[61]

Thus the commanders of the *Ostheer* joined in the general propagandistic effort to paint an inverted picture of the reality on the Eastern Front. These pronouncements were all the more authoritative because they were delivered by men professionally equipped to assess the nature of the war. The gist of the argument was in all cases that the attack against the Soviet Union had merely been a preventive measure, intended to thwart the approaching invasion of "Asiatic barbarism," led by "Judeo-Bolshevism," which had aimed at devastating Europe and destroying its "culture." Precisely because the threat was so great, everything was allowed, indeed, everything must be done, to eradicate the Soviet enemy's "power basis"—that is, the communists and the Jews. No amount of evidence to the contrary could undermine this logic, which conveniently shifted the responsibility for the murderous Nazi policies in the East to their victims. Little wonder that throughout the war the Wehrmacht consistently relied on the same argument. In late December 1941 the commander of II Corps reminded his troops that theirs was merely a defensive war against a barbarous enemy:

> The battles of the previous months have shown you that the Russian soldier is . . . prepared to commit any vile act, be it murder or treachery. . . . What would have happened had these Asiatic Mongol hordes succeeded to pour into Europe and particularly into Germany, laying the country waste, plundering, murdering, raping?[62]

Neither the officers nor the rank and file seem to have been particularly disturbed by the inherent contradictions of their propaganda. After all, by early 1943 the commander of 16 Army was only one among many officers who insisted that the Wehrmacht must manifest precisely that kind of fanatic determination previously so condemned as regards the Red Army. On the occasion of the "tenth anniversary of the National Socialist movement's victory," the troops were especially urged to remember that it too had been

> fought for in tough battles, with a fanatic belief and an unshakable confidence in the Führer, [and] had created the preconditions for the victory of arms in the struggle of the German people for a new and just order of existence. . . . United and strong in its belief in the justice of its cause and with an iron will for victory, National Socialist Greater Germany is now in the fourth year of its struggle for *Lebensraum*. Many men have sealed their love to Führer and Reich with death.[63]

Nor should one think that such ideological pronouncements were limited to the public sphere. Just as von Fritsch had revealed his prejudices in a private letter in 1939, so Colonel-General von Richthofen, commander of 4 Airfleet, made the following remarks in his diary as late as January 1943:

> I am reading again the chapter in "[*Mein*] *Kampf*" about Russian and Eastern policies. Still very interesting and provides answers for almost all questions also in the present situation. Will take care to emphasize these arguments more strongly to the troops in the whole area.[64]

Divisional commanders too showed much concern for their men's ideological convictions. One way of ensuring that the troops would receive the appropriate instruction was by conducting weekly political sessions at company level, for which officers were supplied with ample indoctrination material. This practice was begun by the 12th Infantry Division, for instance, in July 1940,[65] and by June

1943 indoctrination was further intensified by the introduction of special "educational officers," specifically charged with the political instruction of the troops. As the divisional commander explained,

> on the eve of the fifth year of the war the significance of a unified instruction . . . of the troops increases. . . . [Therefore, although] the commanders carry the basic responsibility for this work of [ideological] instruction and education . . . [from now on] they will nominate educational officers for their advice and support.[66]

This ideological instruction was generally greeted with enthusiasm by the troops. One of the division's battalions reported:

> Platoon and company commanders, as well as other officers, discussed current political issues. In many bunkers radio-connections were installed, so that music, news broadcasts, and political speeches could be heard. . . . The soldier is thankful for any change . . . [and] manifests an interest in instruction in political and other current issues, which comes to show that he is more preoccupied with them than one usually thinks.[67]

The 18th Panzer also expressed a lively interest in the ideological instruction of its soldiers. This formation introduced "educational officers" already in autumn 1942, and the divisional commander reported that here too they were warmly welcomed by both officers and men:

> The initiative was viewed quite positively by all officers and welcomed enthusiastically by some. The view that owing to the length of the war the mental energy of the men had to be particularly preserved and encouraged, and that this cannot be achieved to a sufficient degree by the conventional means of entertainment, was generally accepted. . . . The soldiers listened to the lectures attentively. In many cases there is an inner response and a need to be spoken to in such a manner. . . . Doubtlessly this institution depends particularly on the personality of the officers nominated to carry out this work; they must have the confidence of the commanders and the unit leaders. Where this is the case, their work is *very* valuable. In the division this institution has till now proven itself and promises good results once it is further expanded.[68]

The GD Division shared the view that ideological instruction was highly important for the morale and motivation of the troops. Upon establishment in spring 1942, the divisional commander pointed out to his company commanders that they were not responsible only "for pure leadership and training," but were also charged with making extensive use of the company instruction sessions and the appropriate propaganda material so as to provide their men with ideological training.[69] In September the division stressed once more that the Wehrmacht could win only "[b]y our unshakable belief that we are and will remain absolutely superior to the enemy even in the most critical situations."[70] And in April 1943 the divisional commander stressed once more that

> The length of the war calls not only for extraordinary efforts regarding the military performance of the Wehrmacht, but also makes demands upon the power of resistance of each individual soldier. This mental power of resistance has to be repeatedly strengthened, particularly during rest periods. This will be achieved by:
> 1. The uniform orientation of commanders and troops in ideological issues.
> 2. Strengthening of soldierly qualities: bravery, toughness, the will to fight and to obey.
> 3. The recognition of the historical significance of the war.
> 4. Creation of a confident view of the military and political situation even in the face of setbacks and the length of the war: education to steadfastness in crises.[71]

The following month the division introduced "educational officers" and charged them with carrying out all propaganda and "political-educational" instruction among the troops.[72]

Toward the end of the war, when the objective situation became evidently hopeless, combat formations intensified their indoctrinational efforts even more in a desperate attempt to make up for their material weakness. In January 1945 the 4th Panzer Division issued to its troops a so-called "Front-Credo" (*Frontbekenntnis*), which compressed the essence of the National Socialist *Weltanschauung* into a pseudo-religious statement of belief:

> I PROFESS—in view of my oath to the flag—my front comradeship to my division.
> I AM DETERMINED to give my whole strength, my blood and my life in the present decisive battle for the life of my people.
> NEVER will I abandon my weapons. . . .

I BELIEVE in Germany. I will also do all in my powers to preserve and to strengthen the spiritual power of resistance of the German people at the front and in the homeland by speech and deed.

I BELIEVE in the German people united by National Socialism and in the victory of its just cause.

I BELIEVE as a National Socialist soldier in my Führer Adolf Hitler.[73]

By the last year of the war, the Wehrmacht began making use of so-called National Socialist Leadership Officers (NSFOs), who even more closely resembled the hated Red Army commissars than the "educational officers."[74] These missionaries of the Nazi cause were particularly active in disseminating concise and unambiguous battle-slogans among the troops. Typical of the numerous statements of belief common in the Wehrmacht during the last months of the war was the following propaganda leaflet issued by one of the divisional NSFOs:

1. Asia has never defeated Europe. We will break the Asiatic tidal wave this time too.
2. A rule of Asiatic subhumans over the West is unnatural and contradicts the sense of history.
3. Behind the flood of the red mobs sneers the distorted face of the Jew. His craving for power will be broken, as it was once broken in Germany.[75]

The evidently tremendous propagandistic efforts made by the Wehrmacht's commanders notwithstanding, many historians are reluctant to accept that the junior officers and rank and file of fighting units could have been motivated by ideological arguments either in combat or in their treatment of enemy prisoners and civilians. Partly, this has to do with the difficulty of associating the hodgepodge of nonsense which paraded itself as the Nazi *Weltanschauung* with the remarkable professionalism of German soldiers; partly, it has to do with structural developments within the discipline. Thus while historians dealing with civilian society pay little attention to the army, military historians are mainly concerned with military matters, touching on the contact between soldiers and civilians only at the higher levels of either hierarchy, in matters concerning strategy, politics, and economy. Consequently the junior ranks of the army are treated as a gray, faceless mass, devoid of both a civilian past and an individual identity, will, and consciousness. This makes it possible to ascribe to this mass conscript army whatever characteristics one chooses without, however, providing much evidence to sustain one's opinion. The Wehrmacht's soldiers may either be presented as fanatic Nazis, or as entirely indifferent to ideology, depending on one's more general view of civilian society in the Third Reich, and with reference neither to the special conditions under which the soldiers lived, nor to the particular background they came from as recruits. Indeed, just as conscripts supposedly shed all their civilian attributes once they are in uniform, so too their experiences as soldiers seem to be completely erased upon their return to civilian society. Needless to say, after years of education in the Third Reich, a young man's civilian background was just as crucial to his conduct as a soldier as the experience of six years in war was crucial to molding his civilian identity upon release from military service. To be sure, the tendency of historians to underestimate the relationship between army and society has itself to do with the abhorrence of things military following the carnage of the war, just as their scepticism toward ideological motivation is a direct consequence of the disillusionment with the ideals of the first half of this century, whose great promises were drowned in rivers of blood. To these more general causes can be added the fact that in the West the Wehrmacht had by and large observed the rules of war as far as POWs and the civilian population were concerned, excluding of course political and "biological" enemies; ideology did play a lesser role in the conquest and occupation of Western Europe.[76] The war in the East was indeed very different. It was the Soviet troops and population who had to bear the main brunt of the war and of Nazi barbarism, just as it was they too who finally broke the Reich's military might. But the Western experience and memory was different, and was thus differently reflected in the writings of Western historians. Not unrelated was also the fact that the liberal tradition of rigidly separating between politics and the military hampered a clearer understanding of the radically different tradition in Germany or, for that matter, also in Russia. Ironically, while the Cold War and the fear of communism made for a view of the Red Army as a highly politicized institution, it also created the basis for the resurrection of German military institutions, which in turn made it politically necessary to repress the notion of Nazi penetration into the army of the Third Reich.[77] At the same time, those components of the Wehrmacht's propaganda which had presented it as the bulwark of civilization against Bolshevism now appeared admirably suitable for the needs of the Western Alliance. But for the fact that this had come about only after the Nazi regime had been destroyed, Goebbels would have certainly had much cause for satisfaction. And, because during the last decade East-West relations have improved, the growing distance of time between the Third Reich and present concerns seems to justify an approach according to which the Wehrmacht's soldiers should also be considered as the victims, rather than the instruments of Hitler's regime. Thus another distortion of the reality of the Wehrmacht's war on the Eastern Front was gradually legitimized.[78]

Nazi indoctrination in fact had a major and insufficiently acknowledged impact on the perception of reality of all ranks in the German army during the war, and its effects can be seen to have lingered on for many years after the "capitulation." The degree to which the regime's "world-view" penetrated into the minds of its soldiers does not call for much reading between the lines, for a wide array of evidence is readily available. Similarly, examples of the manner in which Nazi arguments were still being used to justify the Wehrmacht's actions long after the war was over are not difficult to come by. Note, for instance, the published memoirs of the Panzer general Heinz Guderian. Whereas regarding the army's involvement in the implementation of criminal policies he takes a familiar apologetic line and simply falsifies the evidence, Guderian expresses full agreement with the Führer's contention that the invasion of Russia was merely an attempt to save the Reich from "being overwhelmed by the Asiatic Bolshevik flood... from the east." Indeed, to his mind "Barbarossa" was a noble struggle whose goal was to defend "European civilization." Hitler, he maintains,

> was clearly aware of the threat that the Soviet Union and the communist urge to world hegemony offered both to Europe and to Western civilization. He knew that in this matter he was in agreement with the majority of his fellow-countrymen and, indeed, with many good Europeans in other lands.[79]

This was written in 1952, when many "good" Europeans and Americans were certainly anti-communist, though far fewer would have been flattered by Guderian's association of this position with Nazism. Interestingly, this view of the Wehrmacht as having played the role of Europe's savior from communism sheds a different light on the argument made by German generals both during and after the war, that any opposition to the regime would have in any case been impossible due to the support Hitler enjoyed among the junior ranks of the army. Guderian writes that "as one year succeeded the next" following the Nazi "seizure of power,"

> the opposition within the Army was continually weakened, since the new age groups that were now called to the colors had already served in the Hitler Youth, and in the National Labor Service or the Party, and had thus already sworn allegiance to Hitler. The Corps of Officers, too, became year by year more impregnated with young National Socialists.[80]

This was obviously true. But what Guderian fails to mention is that he himself featured prominently among Hitler's enthusiastic supporters, not merely because the Führer had made possible the realization of his technological dreams by rapidly promoting him to a position of influence, but also since he became convinced in the most fundamental tenets of National Socialism, namely, the need for expansion of German power and the consequent necessity totally to destroy "Judeo-Bolshevism" or, in its postwar version, "Asiatic barbarism" and communism. The difference between the junior ranks and the Wehrmacht's young generals was thus not as great as such memoirs try to make out; both had in common a favorable opinion of the Nazi *Weltanschauung* as they understood it, and a willingness to implement those aspects of it which were relevant to their functions as soldiers. Thus Guderian is really also referring to himself and to his peers when he writes:

> When National Socialism, with its new, nationalistic slogans, appeared upon the scene the younger elements of the Officer Corps were soon inflamed by the patriotic theories propounded by Hitler and his followers.[81]

If we understand that what many of these officers believed to be "patriotic theories" were actually National Socialist notions of geographical expansion and "racial" destruction, we can view the above passage as a candid statement of the extent to which the regime had succeeded to implant its world-view into the minds of its soldiers.

Such pronouncements by a former Wehrmacht general merely eight years after the end of the war may not be particularly surprising, though the tendency of Western scholars and soldiers to go on considering such men as mere professionals certainly is. The influential British military historian Liddell Hart was among those who set the tone for this approach, when he wrote soon after the fighting ended how much he had been taken by the supposed "gentlemanliness" in war of the German generals he had spoken with.[82] An RAF officer who contributed an introduction to a book published originally in 1952 by the German pilot Hans-Ulrich Rudel claimed that "although I have only met him for a couple of days he is, by any standard, a gallant chap and I wish him luck."[83] Brader did not seem to think that Rudel's gallantry was somewhat qualified by his obviously intense loyalty to Hitler and everything he had stood for, not only until the very end of the war, but even while writing his autobiography. Indeed, this tough Stuka pilot's outburst of emotion upon hearing of Hitler's suicide is accompanied by a sense of continuing mission quite unencumbered by his Führer's fate. Thus, while the suicide is said to have had "a stunning effect upon the troops," Rudel feels that as

> the Red hordes are devastating our country . . . we must fight on. We shall only lay down our arms when our leaders give the order. This is our plain duty according to our military oath, it is our plain duty in view of the terrible fate which threatens us if we surrender unconditionally as the enemy insists. It is our plain duty also to destiny which has placed us geographically in the heart of Europe and which we have obeyed for centuries: to be the bulwark of Europe against the East. Whether or not Europe understands or likes the role which fate has thrust upon us, or whether her attitude is one of fatal indifference or even of hostility, does not alter by one iota our European duty. We are determined to hold our heads up high when the history of our continent, and particularly of the dangerous times ahead, is written.[84]

Rudel's views of the war had thus been distorted to such a degree by Nazi indoctrination that he stuck to them long after the fighting was over, and had no fear of publishing them either in Germany or in an English translation. Nor did he have much reason to worry, as the warm introduction by an RAF officer testifies. Rudel describes the fate and role of Hitler's regime in Europe using precisely the same propagandistic terms so prevalent during the war. The Third Reich is not the terrible destroyer of human beings and moral values, but their defender; the German soldier is not Hitler's instrument of genocide, but a sort of Germanic St. George spearing the communist dragon. Postwar knowledge of the real essence of the Nazi dictatorship seems to have had little impact on such men, who in any case must have known a great deal about the regime's murderous nature long before it was finally destroyed.

Rudel's pronouncements, which are fairly representative of the memoir literature of the first two decades following the end of the war, are a good indication of the manner in which men whose minds had been permeated by Nazi ideas retained these views in spite of the obvious consequences of Hitler's policies. But the Nazi *Weltanschauung* had a considerable effect on some of the regime's domestic opponents as well, reflecting as it did in a more radical form some of the aspirations and hopes of German nationalism as it had been molded at least since Bismarck. This in turn has meant that some of the key arguments and terms employed by the regime during the war have recently resurfaced in the Federal Republic under the guise of an attempt to give the Germans back their history and allow them to regain their national identity by recognizing the positive aspects of even the murkiest periods in their past. Most disturbingly, ideological, geopolitical, and nationalist justifications for the role played by the Third Reich, and especially by the Wehrmacht, in "stemming the Bolshevik flood," have appeared in works by several eminent and respectable scholars, thereby legitimizing a historical interpretation which consciously or not credited Nazism with the same sort of achievements it had claimed for itself in the last phases of the war.[85] The most pertinent example of this scholarly distortion of the past in the name of the future as regards the Wehrmacht was a little book published in 1986 by Andreas Hillgruber.[86] To his mind, when observing the "winter-catastrophe of 1944/45," that is, the Red Army's penetration into East Prussia, the (apparently German) historian

> must identify with the concrete fate of the German population in the East and with the desperate and costly [*opferreichen*] efforts of the German *Ostheer* and the German navy in the Baltic Sea, which sought to defend the population of the German East from the orgy

of revenge of the Red Army, from the mass rapes, the arbitrary murders and the innumerable deportations, and to keep open the flight routes to the West over land or sea for the East Germans in the very last phase [of the war].[87]

The Russians, maintains Hillgruber, had committed precisely the sort of barbarities Nazi propaganda had always predicted they would. Consequently, the historian ought to have no qualms about choosing the German victims of the Bolshevik "orgy of revenge" as his object of Rankean empathy. The author is well aware that the Russians' desire for revenge was a direct result of the horrors inflicted upon them by German occupation; he similarly admits that as long as the Wehrmacht kept fighting in the East, the extermination of the Jews could continue undisturbed; yet he finds his choice quite natural and unproblematic, and rejects the idea of empathizing either with the regime's victims, or with its domestic and foreign opposers. Thus Hillgruber perpetuates a kind of continuity quite different from that actually meant by the "revisionist" historians. In 1941 Wehrmacht propaganda warned that had Germany not invaded Russia, it would have been subject to a barbarian invasion from the East; the German soldier was therefore ordered to do what otherwise would have allegedly been done to him. In 1944 it was claimed that these predictions were coming true; hence the Western Allies were asked to join hands with the Nazi regime against the "Asiatic flood." This argumentation was also adopted in the apologetic memoir literature of the 1950s and the 1960s. Finally, in 1986 a respectable German scholar could claim that the *Ostheer*'s battles of 1944–45 had been a heroic effort to stem that same "flood" and should consequently be seen as a glorious chapter in German history, even if this chapter had been written during the Nazi regime's most frenzied period.

Hillgruber does not dwell on the fact that the Red Army was actually far more lenient to the occupied Germans than the Wehrmacht had been to the Russians, and that although the Soviet Union established ruthless dictatorships in the East, it did not have the same genocidal intentions partly carried out by the Third Reich. After all, had this not been the case, German reunification would have remained a mere theoretical issue. Quite on the contrary, this author speaks of the "Soviet view of war, which had clearly by and

large taken on such barbarous characteristics during Stalin's epoch."[88] Not only does he confuse chronology, lacking to mention
that the Red Army came to Berlin only as a reaction to the Wehrmacht's almost successful attempt to conquer Moscow, he also tries
to show that there was no fundamental difference between Nazi,
Soviet, and Western barbarism. Once more repeating the propagandistic claims of the period, Hillgruber maintains that the Western
Allies refused to rescue the Reich from the Russians merely because
by achieving the total destruction of Germany, their own position
would have remained unchallenged. Indeed, he believes that their
sinister plans were the legacy of a general European, rather than a
specifically Nazi manner of thinking:

> The widespread idea in Germany during the First World War of a
> racial field- and floor purge, which was carried out by the German
> and Soviet sides since the beginning of the Second World war in
> September 1939, was now—without it being possible to find direct
> links to the expulsions of portions of the population in the East—
> also introduced by the British side as an element in their own war
> plans, as it appeared likely to promise long-term security for their
> own role of leadership in Europe.[89]

The Germans might have thought of genocide first, and might have
practiced it along with the Russians, but the British were apparently
not far behind, in theory if not in practice.

Conversely, Hillgruber finds that not only the circumstances of
the war in the East, or the genocidal plans of Germany's enemies,
but also some central aspects of the Nazi regime's own policies,
make it possible to identify with the Third Reich and lament its
total demise. As he points out,

> the Eastern political concept of the liberal-conservative opposition
> deserves a just appreciation; it was the only active alternative in
> Germany to Hitler's radical utopia. Common to both conceptions,
> that of Hitler and that of the liberal-conservative opposition, was
> only [!] the conviction that Europe had to be organized and led,
> or in Hitler's view, ruled, from the middle, by the German Reich.[90]

With this aspiration of both Hitler and his (liberal?)-conservative
opponents Hillgruber wholeheartedly agrees.[91] Indeed, in his view
it was not only a German, but also a European tragedy that a
Mitteleuropa "organized and led" by Germany had failed to come

into being due to the machinations of the flanking superpowers which thus ensured Europe's political impotence. This brings him to the conclusion that the German army was engaged in a fight for two noble goals, precisely those, incidentally, ascribed it by Nazi propaganda. First, writes Hillgruber,

> the German *Ostheer* defended in an entirely elementary sense the people of precisely these Prusso-German Eastern Provinces, who were threatened by the gruesome fate of having their homeland flooded by the Red Army. . . . The oft-repeated claim of National Socialist propaganda, that there was no alternative between Hitler and Stalin, now became a reality for the Germans in the East.[92]

Hillgruber readily adopts this perception of reality for, along with many postwar apologists, he accepts the central contention of the Nazi regime that bad as Hitler might have been, he was by far better than Stalin. Consequently, one was justified to defend the Nazi regime of the former from the "Bolshevism" of the latter, especially as there was ostensibly no other alternative. Nazi propaganda had been right all along, as were the Wehrmacht's soldiers. Moreover, as we learn from Hillgruber's second conclusion, this was not merely a battle for survival, but also a fight to retain Germany's greatness. In these last months of the war, he writes,

> the German *Ostheer* also struggled—only partly informed of the Allies' war aims by the half-truths of the National Socialist propaganda—in desperate defensive battles for the preservation of the independence of the German Reich's great power position, which according to the will of the Allies was to be smashed.[93]

The ultimate defeat of the Wehrmacht inevitably spelled the defeat of Europe, whose heart was torn out and whose body was left at the mercy of the superpowers in the periphery. Put differently, both in the "elementary sense" of fighting for their own and the civilian population's survival in the face of a barbaric invasion, and in the politico-strategic sense of defending Europe from the domination of non-European powers, the troops of the *Ostheer* were, as Nazi propaganda had claimed all along, fighting for a just cause. Hillgruber thus provides us with a disturbing example of the manner in which merely forty years after the event a scholarly aura can be given to a distorted interpretation of the Third Reich's war based to a large degree on the regime's own propaganda.[94] The only re-deeming feature of this presentation, and it is an important one in this context, is that it supplies us with powerful proof of the lingering effects Nazi views have had not only on veterans, but also on schol-ars. In another sense, Hillgruber has defeated his own ends, for he has taken for granted what most historians have been reluctant to admit, namely, that rather than being a victim of the regime, the German soldier of the Second World War was a fighter by conviction for what he had been taught to believe were existential and moral goals. Hillgruber does not provide much documentation for this assertion, but as we shall see below, the Wehrmacht's troops did indeed come to accept the Nazi view of reality. Indeed, it was this belief in the regime's propaganda which kept them fighting even when their units disintegrated and military discipline broke down. This does not mean that every individual German soldier was a committed National Socialist; rather, it is to say that the vast ma-jority of the troops internalized the distorted Nazi presentation of reality, and consequently felt that they had no other alternative but to fight to the death.[95]

Among the Third Reich's higher military and political echelons there was little doubt that the rank and file and junior officer corps were highly committed to the regime. This view was shared both by the minister of propaganda and by the conspirators, by officers seeking an excuse to stay out of trouble and continue collaborating with the regime and by those who hoped to overcome the military crisis by strengthening the will and determination of the soldiers. Studies of morale in the Third Reich have stressed that until late in the war combat troops remained in higher spirits than the population in the rear, their firsthand knowledge of military setbacks notwith-standing. One historian has claimed that "periods of buoyancy were triggered mostly by the confidence and attitude of the front-line soldiers," who were "the staunchest supporters of Hitler and the regime." The authorities were well aware of this, and by 1943 "[m]obilization of officers and soldiers to raise the public mood . . . had long been introduced."[96] SD reports in June 1943 led another historian to conclude: "The 'Führer myth' remained relatively strong . . . [among] ordinary soldiers."[97] Following the attempted assassination of Hitler in July 1944, another "morale report" main-tained that today [people] think that for some time the traitors have sabotaged the Führer's objectives and orders. This opinion is primarily due to an increase in the written and oral reports by soldiers from the Eastern Front who declare that they are now discovering the reasons for the absence of reinforcements and the often senseless shifting of units and exposure of the front.[98]

American surveys among German POWs revealed that more than two-thirds of the soldiers expressed "belief" in the Führer in August and late November 1944.[99] An army report from mid-December 1944 pointed out that there was little defeatist talk among the troops, and that "[t]here is a firm conviction that the tremendous military efforts of our people will lead us to victory."[100] In July 1941 Goebbels wrote in his diary that "our soldiers at the [Eastern] front are now completely convinced of the necessity of this war,"[101] and added a few days later: "Morale of our men at the front [is] very good. The soldiers now realize that this campaign was necessary."[102] If this was perhaps to be expected at a time of great victories, as late as March 1945 Goebbels insisted that according to Allied sources "our men have been fighting like savage fanatics,"[103] and that the troops were "resisting at all costs—to the extent that the situation and their equipment permit."[104] Highly impressed by a visit to one of the combat units, Goebbels wrote that "there is not the smallest sign of defeatism here,"[105] and proudly observed "that the faith in victory and in the Führer is prevalent among these men."[106] Even the prisoners, he went on to say, "still maintain the view that Germany must definitely win the war," as they "have an almost mystical faith in Hitler. This is the reason," he concluded, "why we are still on our feet and fighting."[107]

The same generals who made such efforts to motivate their troops by large doses of National Socialist propaganda also claimed that precisely the success of this indoctrination had hampered any action against Hitler. Thus Manstein maintained in his memoirs:

The preconditions for a *coup d'état* would have been... the following of the whole Wehrmacht and the agreement of the majority of the population. Both did not exist during the years of peace in the Third Reich as well as during the war (with the exception perhaps of the very last months).[108]

Which is, of course, another way of saying that the majority supported Hitler. Though such apologetic claims, found in numerous other memoirs,[109] should not to be taken at face value, it is well worth considering that even those who did find the courage to plot against Hitler were evidently much disheartened to discover that virtually no military units existed which could knowingly be deployed in a *Putsch* attempt. As Johnnie von Herwarth wrote, "the soldiers... were naturally under the influence of Nazi propaganda."[110] Consequently,

> [i]t would have been difficult in any circumstances to identify among the tens of thousands of troops those upon whom we could count. The task of locating them became more vexatious as we realized that few, if any, were likely to fit that category.... We never had any troops upon which we could rely one hundred per cent.[111]

Indeed, as von Herwarth explains, the very decision to assassinate Hitler rather than arrest and try him as had previously been suggested was based on

> the general conviction that German troops would never be willing to accept a different command as long as Hitler lived, but that news of his death would instantly bring about the collapse of the myth that surrounded his name. Hence there was no way of gaining the support of large numbers of German troops without eliminating Hitler.[112]

Postwar accounts which speak of the soldiers' pure professionalism and ideological indifference, whether written by the very same generals who excuse themselves for inaction against the regime by saying precisely the opposite, or by more disinterested historians who, however, often rely on claims made by the former rather than on less biased evidence, can thus not be seen as reflecting the objective reality at the front. One must read the following passage by the German historian Hans Mommsen to understand the degree to which even the most distinguished scholars of the Third Reich tend to reject the notion that Germany's national army of conscripts had become Nazified, all available documentation notwithstanding:

> The picture drawn by the regime's propaganda of troops fanatically fighting for the National Socialist cause, was false even concerning the elite formations of the *Waffen*-SS. ... The mentality of the av-

erage *Landser* was characterized by soberness, rejection of the far-from-reality propaganda tirades, and by a firm will personally to survive. Certainly, under cover [*unter dem Vorzeichen*] of the commissar order there were grave encroachments by the army against the defenseless civilian population handed over to it and against prisoners of war; the partisan war led to an unprecedented brutalization of the conduct of war by both sides [!]. But the average soldier had little influence on this and could hardly find a way of avoiding the escalation of violence.[113]

To be sure, the soldiers were indeed concerned both with their survival and with the various professional aspects of fighting, but rather than diminishing their need for propagandistic reinforcement, the worsening battle situation generally enhanced it. Naturally, it would be false to describe the troops' "mystical belief in the Führer" as a firm commitment to an articulate and coherent ideology; Nazism never made such claims to begin with. Quite on the contrary, as we have noted before, while being contemptuous of the "intellectualism" of rival ideologies, it was highly consistent in presenting itself as an intense emotional state derived from "natural instincts" and "unshakable beliefs." The single most important element in this state of mind was "blind faith" in the Führer. Creating and preserving this unquestioning loyalty to Hitler among the troops played a major role in distorting their perception of reality, both by sustaining their hope in the *Endsieg* and by shifting the responsibility for their own crimes to their victims. The profound impact of indoctrination and propaganda on the soldiers' psyche can clearly be seen from their private correspondence.

The most striking aspect of the soldiers' letters is the remarkable similarity between their terminology, modes of expression, and arguments and those which characterize the Wehrmacht's propaganda. In complete contradiction to Mommsen's above quoted assertion, the fact that these men, who were indeed closer than any of the propagandists to the reality of the war, saw and described it through the distorting lenses of the regime's ideology, is the true measure of the extent to which they had been made into Hitler's soldiers in the most profound sense of the term—that is, that they perceived reality at the front just as he did in the safety of his bunker, sharing his fantasies of conquest and grandeur, of racial genocide and Germanic world rule. In fact, even those soldiers who expressed criticism of the regime were infected by the Nazi vocabulary. Nor should we find this particularly astonishing for, as some scholars have powerfully shown, the National Socialist perversion of the German language was so profound, that it affected even the regime's victims.[114] This phenomenon too has had a lingering effect on German society, since efforts to "purge" the language after the war have not been altogether successful, both because the Nazis had already "purged" it once before from "foreign" influences (and some German dictionaries still have special volumes of *Fremdworte*), and because the vast literature on Nazism has had the paradoxical effect of keeping Nazi terms in constant circulation. Seen from this perspective, it was inevitable that after years of premilitary and army indoctrination the Wehrmacht's troops would be able to assess and describe reality only by constant reference to the Nazi *Weltanschauung*, which in their case literally constituted their view of the world. Considering its consequences, the creation of this consensus among the troops was probably the single most significant achievement of the Nazi regime's educational efforts.

As indoctrination greatly intensified during the Russian campaign, and as prejudices against Slavs and Mongols, Gypsies and Jews, and of course Bolsheviks, were far greater than against the inhabitants and political systems of Western Europe, ideological references and racist sentiments were far more prevalent in letters from the Eastern Front. But long before "Barbarossa" the soldiers' correspondence reflected the extent to which they had come under the influence of the regime's *Weltanschauung*. As early as 3 September 1939 the twenty-four-year-old Heinz Küchler was convinced that he was fighting "a new battle for a better future," and that

> this time the decision must be reached, whether the chaotic circumstances should result in the self-dissolution of our culture or in a new rational world-order. . . . We should have no fear of this battle. . . . [W]e should also not value our lives too dearly. . . . Our greatness must lie in the ability, not to master fate, but to preserve our fate in spite of our personality, our will, our love, and unreservedly to sacrifice ourselves to a world-order which is not our own. Even the frightful plight, perhaps it precisely, will bring us closer to cognition and truth.[115]

Similarly, on 9 February 1940 Karl Fuchs, then serving with the 7th Panzer Division, wrote to his father:

> We all hope that we'll be transferred to the front soon. That is our greatest wish!... [I]n the final analysis, this must be the highest and most noble goal: a man must prove himself in battle. This battle is not only an individual struggle but also a struggle for our family as well as our German people.[116]

The great victories in the West led to increasing adulation of Hitler and consequently to a growing agreement with his "ideas." This can be seen, for instance, from the intelligence reports of 16 Army during its occupation of France and the Netherlands in the second half of 1940 and early 1941, which included an evaluation of the troops' morale based on an analysis of their letters. In November 1940 it was reported that the men's letters "express very frequently a general confidence and a belief in the rightness of the Führer's policies."[117] Thus one soldier wrote:

> We Sudeten-German SA men want to be in the very front-line, in order to render a fraction of our thanks for the liberation of our beautiful homeland by our magnificent Führer Adolf Hitler. In this struggle too our Führer will lead us victoriously.

Another soldier expressed the same kind of determination to fight on:

> Yes, we are merely front-soldiers, who will remain in this situation and never rest, till Germany is completely victorious. Yes, the homeland should be filled with new pride of her sons, who will still knock into the ground even the last enemy [England]. No sacrifice will be too great for us.—Be strong, and bear with patience all that destiny will demand of you.

A particularly interesting letter repeated almost word for word what could be read in most propaganda sheets of the period, affirming that the troops had come up against precisely that kind of "reality" they had been led to expect, and finding this as further proof of the superiority of the Reich's political system and of the Germans' obvious superiority:

> We are all burning to be allowed to present those who are guilty of this great war [England and the Jewish "plutocrats"] with the

last reckoning. And the reckoning will be made precisely, this we have sworn to our dead comrades.... Now we are stationed in France and have had more than enough of the moral, ethical decay, which appears to us here again and again. We repeatedly recognize that this people can hope for no salvation from the mire into which it has sunk. There one can see for the first time how beautiful Germany is, and how proud we should be of being German, and thankful to our Führer, who had spared us from the misery which we now see daily.[118]

This view was echoed by many other soldiers. Heinz Küchler wrote as early as 3 June: "It is still unclear to me what had made the rapid German victory possible," and then added in English: "There was certainly something wrong in the state of France!"[119] Karl Fuchs wrote his new bride on 15 July:

Later my soldier friends and I went to a "bookstore" in Versailles. You can't imagine what junk and pornography we saw!... You can truly see that in the areas of cleanliness and morality, the French people have skidded to a new low. Such an incident is simply unthinkable and impossible in our German Fatherland. When a society is capable of reducing feminine beauty to such a level, then this society has lost its right to be called a "grande nation." Yes, this society has lost not only its vitality but also its morality.[120]

In January 1941 16 Army's intelligence report once more expressed satisfaction with the troops' confidence in the Führer and in the ostensibly fast approaching final victory. The report noted that

Many letters can serve for the strengthening of the home-front on account of their content. They fortify the sense of attachment between the homeland and the front and serve as a guarantee that among the troops a good soldierly bearing and a decent German conviction always have the upper hand.[121]

What the intelligence section meant by this "decent German conviction" was naturally a complete trust in Hitler and the Wehrmacht's commanders. And indeed, the soldiers' letters showed that they both received and read the propaganda material amply supplied to them, for their interpretations of the war situation were undeniably similar to the regime's official line. Thus one soldier wrote that "John Bull will certainly not survive a second summer, the

Führer and his proud Wehrmacht will take care of that. I am proud to have actively helped at the front in the army."[122] Another man noted: "We are working on giving England the last blow and then there will be calm. Then the great peace will come for which all peoples are hoping. Fighting for that, no sacrifice is too great." This soldier obviously believed that he too was fighting a war to end all wars, and that a Europe ruled by Hitler would revert to peace and prosperity, as the regime repeatedly claimed. Others were less peacefully minded. One letter saw in the coming battle an opportunity for revenge and a heroic death, and expressed the kind of total trust in Hitler and quasi-religious fatalism which Nazism had done so much to promote among its followers:

> This time it is really impossible for me to celebrate Christmas at home. But that is of no harm; because the Fatherland and the Führer have called us, and we have followed their call with a happy heart. Our comrade Fritz Lehmann has brought a much greater sacrifice, he has given his life. He has fulfilled the most beautiful duty of a soldier and has become an example for us all, which we have to follow. We still have to gain what he will no longer be allowed to experience, namely the victory over our craftiest enemy [England]. We will achieve this goal. Our lives belong to God and to the Fatherland, and may they decide over our fate.[123]

Such views were shared by many other soldiers. Thus, for instance, Karl Fuchs wrote his father on 3 August 1940 that "the days are numbered for those bums over there in England. They won't be able to attack German cities and peaceful farms anymore. All of us feel that once we're over there, no one will show any mercy whatever, no matter who's involved."[124] And on 1 September he promised his wife that it "will be a pitiless and dreadful time for England,"[125] and urged her to "think of our Fatherland and our Führer to whom we give everything as his children."[126] Fuchs, like many of his generation, saw his love to his wife and his love to Hitler as a single entity symbolizing the spirit which united the whole German *Volk*. Thus he wrote on 9 November 1940:

> The last words of the Führer's radio address are over and a new strength streams through our veins. It is as if he spoke to each individual, to everyone of us, as if he wanted to give everyone new strength. With loyalty and a sense of duty, we must fight for our

principles and endure to the end. Our Führer represents our united German Fatherland. . . . What we do for him, we do for all of you; what we sacrifice in foreign lands, we sacrifice for our loved ones. When the Führer speaks on these festive occasions, I feel deep in my soul that you at home also feel that we must be ready to make all sacrifices. . . . German victory is as certain as our love for each other. Just as we believe in our love, so we believe in our final victory and in the future of our people and our Fatherland.[127]

And the following day he added in a letter to his father that having heard Hitler's "magnificent and overwhelming" speech, he had come to feel "the greatness and the intensity of our times" and to understand that "only one thing matters and that is our German Fatherland."[128]

The correspondence from the Eastern Front provides us with a particularly good opportunity to observe the manner in which German troops internalized some of the central notions of National Socialism and employed them to rationalize their predicament at the front, legitimize their criminal actions, and fortify their spirits. Naturally, much of what the soldiers wrote was heavily influenced by the Wehrmacht's propaganda. But it is extremely revealing that they incorporated these arguments in their private correspondence, given the fact that censorship was concerned with incidents of criticism, not with the absence of Nazi phraseology. The soldiers' letters reflected the distortion of reality among the troops in two significant spheres: first, the dehumanization and demonization of the enemy on political and racial grounds, with a particular reference to the Jews as the lowest expression of human depravity; and, second, the deification of the Führer as the only hope for Germany's salvation. Intermixed with these central themes were notions regarding battle as a supreme test of character and manhood, as well as of racial and cultural superiority, and a view of the war as a holy crusade for a better future and against an infernal host of enemies sanctioned by God, who among the more pious and philosophically inclined at least partially replaced Hitler as the arbiter of German and universal destiny. On the first day of the campaign Hermann Stracke, a twenty-two-year-old student of philosophy, wrote: "Last night the lieutenant read us the Führer's appeal, and now we may at last take part." This would be "the war of Young Europe against the greater part of Asia," and though "the will of destiny remains obscure," he insisted that "I go to the battle confidently, glad and undaunted, and proudly take upon myself this test of life."[129] One sergeant wrote that he had always known "that in the long run no friendly relations can be maintained with the Bolsheviks," especially as "there are far too many Jews there."[130] Two days later another NCO explained that "Jewry has declared war against us along the whole line," and added: "The Marxists fight shoulder to shoulder with high finance as before 1933 in Germany." Yet he was reassured by the knowledge that "in Germany National Socialism had won."[131]

As the campaign progressed, the resistance of the Red Army and increasingly also of the civilian population seemed to confirm what they had been told, while the atrocities committed by the Wehrmacht and the SS were attributed to the enemy's malicious character rather than the murderous policies of the Nazi regime. Indeed, the inversion of reality soon became a consistent component of the correspondence from the front. On 28 June Karl Fuchs wrote his wife and baby son that the Russians "fight like hired hands—not like soldiers, no matter if they are men, women or children on the front lines. They're all no better than a bunch of scoundrels." He hasten to reassure them, however, that "Europe stands under the leadership of our beloved Führer Adolph [sic] Hitler, and he'll re-shape it for a better future."[132] Another soldier promised in early July: "This time an end will certainly be put to this God-hating power," and expressed his shock at observing "evidence of Jewish, Bolshevik atrocities, the likes of which I have hardly believed possible.... You can well imagine," he went on to say, "that this cries for revenge, which we certainly also take."[133] A Wehrmacht major who happened to pass through Warsaw on his way to the front went so far as to ascribe the horrors of the ghetto not to Nazi barbarism, but rather to the Jews' own inhumanity:

> The conditions in the ghetto can hardly be described.... The Jew does business here with the others also on the street. In the morning, as I drove through in my car, I saw numerous corpses, among them those of children, covered anyhow with paper weighed down with stones. The other Jews pass by them indifferently, the primitive "corpse-carts" come and take away these "remainders" with which no more business can be done. The ghetto is blocked by walls, barbed-wire, and so forth.... Dirt, stench and noise are the main signs of the ghetto.[134]

Yet it never occurred to this officer that these Jews were being starved to death by his fellow-countrymen. Similarly, another man reported from Minsk in October that

> Following [partisan] attacks a number of people, especially Jews, are summarily taken and shot on the spot, their houses set on fire. Recently . . . POWs were simply shot down in a pile by the guards. The lads quarreled over bread and old pieces of clothing which had been thrown to them. Three lay dead and were instantly buried on the spot by the Jews like mad dogs.[135]

This man reacted in the same way to what he saw as others in the rear reacted to propaganda photographs and films; reduced to a state of terrible wretchedness and starvation, the Reich's alleged biological and political enemies came to resemble the stereotype of the *Untermensch* promulgated by the regime. Killing them was consequently no worse than shooting a mad dog. But as the Soviet subhumans were found to resist the Wehrmacht with great ferocity, the soldiers followed the propagandistic line which attributed this unexpected turn of events to the ideological incitement of the Red Army by Jewish Bolsheviks as well as to the generally savage nature of the Russian race. The fact that the Wehrmacht reacted to Soviet resistance by calls for a further "fanaticization" of its own troops by means of intensified indoctrination, and demanded even more "ruthlessness" than previously, merely justified the claim that this was indeed a "war of ideologies" where everything was allowed to achieve final victory. In early August one soldier wrote that "the Russian is a very tough warrior," and explained that "the brutality which constantly characterizes the Russian can be explained only by incitement." Indeed, he believed that the Russians were "a people which needs long and good schooling in order to become human."[136] Karl Fuchs wrote on 5 July that "Russia is nothing but misery, poverty, and depravity!"[137] Two weeks later he promised: "When I go back I will tell you endless horror stories about Russia. Yesterday, for instance, we saw our first women soldiers. . . . And these pigs fired on our decent German soldiers from ambush positions."[138] In early August he reported that the "pitiful hordes on the other side are nothing but felons who are driven by alcohol and the threat of pistols pointed at their heads. . . . There is no troop morale and they are at best cannon fodder. . . . They are nothing but a bunch of assholes!" He thus concluded: "Everyone, even the last doubter, knows today that the battle against these subhumans, who've been whipped into a frenzy by the Jews, was not only necessary but came in the nick of time. Our Führer has saved Europe from certain chaos."[139] Finally in September he informed his mother that "Russia is like a pigsty," and that the Soviet troops were "desperate and right now are driven toward the front with threats from their political commissars. Their last hour has come!"[140] Another soldier was certain that the Red Army's determination was due to the fact that its troops were all "confirmed communists."[141] Indicative of this inversion was a letter by a captain who must have known of the wide-scale maltreatment of Soviet POWs, and yet maintained that "[t]he Russians had been completely stultified and persuaded that the Germans would massacre all prisoners,"[142] implying that they had been lied to and that nothing of the sort was actually taking place.

The refusal of the Russians to capitulate in spite of their alleged inferiority was put to use by the propagandists as a justification for the invasion of the Soviet Union, for it was supposed to prove that the Red Army had secretly prepared to attack the Reich and, that had this plan not been foiled at the last moment, it would have totally devastated Germany. This line of thinking can still be found in some historical writings today, evidence of another heritage of Nazi propaganda, rather than of any documentary proof: "Barbarossa" justified as a preventive war, if not indeed a pre-emptive attack, and German criminal policies qualified by Soviet intentions toward the Germans and actual practice toward their own alleged domestic enemies.[143] The idea of the Gulags as the "originators" of Auschwitz is anything but a recent invention;[144] not only was this a common propaganda claim, it was also widely accepted by the troops. In mid-July 1941 Private Fred Fallnbigl wrote from the front:

> Now I know what war really means. But I also know that we had been forced into the war against the Soviet Union. For God have mercy on us, had we waited, or had these beasts come to us. For them even the most horrible death is still too good. I am glad that I can be here to put an end to this genocidal system.[145]

Lance-Corporal J. F. wrote on 3 August that "to those in the homeland we soldiers can only say that he [Hitler] has saved Germany and thereby the whole of Europe from the Red Army by his decision. The battle is hard," he admitted, "but we know what we are fighting for, and with confidence in the Führer we will achieve victory."[146] Another soldier wrote from Russia in late August 1941:

> Precisely now one recognizes perfectly what would have happened to our wives and children had these Russian hordes . . . succeeded in penetrating into our Fatherland. I have had the opportunity here to . . . observe these uncultivated, multi-raced men. Thank God they have been thwarted from plundering and pillaging our homeland.[147]

Private Kurt Christmann exclaimed a few days later: "What would have happened to cultural Europe, had these sons of the Steppe, poisoned and drunk with a destructive poison, these incited subhumans, invaded our beautiful Germany? Endlessly we thank our Führer, with love and loyalty, the saviour and the historical figure."[148] On 1 September Lance-Corporal O. Rentzsch expressed similar sentiments:

> It is good to know that this confrontation has already come. If otherwise those hordes had invaded our land, that would have . . . made for great bloodshed. No, now we want to shoulder ourselves all endeavours, in order to eradicate this universal plague.[149]

Private Albert Stahl agreed that "Germany can only be glad to have a Führer who is putting an end to this whole spectre. Even if already thousands of brave German soldiers rest here in the earth, then this sacrifice is not in vain, for an invasion of our homeland would have been an end of everything. In this strong belief," he concluded, "the soldier fights with greater courage than ever before and in spite of all dangers."[150] Corporal Alois Hein likewise found it "impossible to contemplate, what would have happened had these beasts come to Germany." Consequently: "Everyone must willingly hold out and make sacrifices to the extreme, for compared to what would otherwise happen, this is nothing."[151] Yet another NCO shuddered at the thought of the terrible fate from which Germany had been saved at the last moment: "Had these cannibalized heaps of soldiers fallen upon Germany, everything which is German would have been done with."[152] And in November Karl Fuchs wrote his mother: "You at home must always keep in mind what would have happened if these hordes had overrun our Fatherland. The horror of this is unthinkable!"[153]

Quite apart from their deep-seated racism and anxiety, the soldiers' letters also betray a need to justify the criminal actions of the army in the East, observed by all and committed by many of the troops. There is little doubt this was largely achieved by dehumanizing the Russians and ascribing to them atrocities they might have committed. The backward conditions in which the village population lived, made even worse by the effects of the fighting, further encouraged the soldiers to think of the Russians as subhumans, while simultaneously reconfirming the troops' belief in the superiority of their own culture, race, and leadership. One NCO wrote: "Our propaganda has certainly not exaggerated, perhaps understated" what he called the "pre-flood conditions" reigning in the Soviet Union.[154] Karl Fuchs noted: "No matter where you look, there is nothing but dirty, filthy block houses. You can't find a trace of culture anywhere. Now we realize what our great German Fatherland has given its children. There exists only one Germany in the entire world."[155] This also made the troops feel that they were actually liberating the Russian people from "Bolshevism, the world's enemy, which had made the Russians into its mercenaries." The Russians' suffering was therefore blamed not on the Wehrmacht, but on the "Bolsheviks," who "do not care whether the Russian people is bled to death," and "have no sense of responsibility."[156] Lieutenant Otto Deissenrath wrote on 30 July 1941: "Everywhere the ghost of Bolshevism stares at us, from the tormented eyes of the peasants, from the dull gaze of the prisoners, from the hundreds of murdered people, from the farm houses, the impoverished villages and the collapsing homes, it often seems to me that this is the devil's work." All of which led him to conclude that he was waging "a battle against slavery, against Bolshevik madness."[157] Private Walter Sperath expressed indignation at discovering that even "animals are treated better back home than the manner in which these people are housed and fed," and he too swore that he and his comrades "will not end this battle before this rabble is eradicated root and branch with the blessing of European culture and humanity."[158] Similarly, Karl Fuchs wrote his mother on 15 October: "Our duty has been to fight and free the world from this Communist disease. One day, many years from now, the world will thank the Germans and our beloved Führer for our victories here in Russia. Those of us who took part in this liberation battle can look back on those days with pride and infinite joy."[159] And a few days later he wrote his wife: "Now that we have been here for some time and have had a chance to become acquainted with this land, we all of a sudden understand why it was an easy thing for the Communist agitators to systematically poison these people." Indeed, he asked her to keep his letters so that "I will remember from them the true character of this country and its people who have been so depraved by these idiotic Communist ideas."[160]

Yet at the same time the racist prejudices of the troops made many of them wholly indifferent to the fate of their victims. In mid-September one soldier seemed to be quite unmoved by the fact that the prisoners his unit had taken consisted mainly of "old men over 40 . . . [and] young, 15-year-old women," especially as they were all "Mongols, Chinese, Asiatics, a mixture in the real sense of the word." This obvious racial inferiority of the enemy must have also been the reason for his satisfaction at observing the numerous hangings of "those who have stolen military property or soldiers who have been roving about in the forests dressed as civilians and who have committed acts of terrors. They remain hanging two or three days as a warning."[161] Lance-Corporal G. S. noted that among "this mixture of races the devil would feel at home. It is, I believe, the most depraved and filthiest [people] living on God's earth."[162] Another sergeant commented with the same matter-of-fact tone that "in response" to the shooting of a German officer by a Russian civilian "the whole village was burned down," then observed that "this Eastern campaign is greatly different from the Western campaign."[163] In mid-October an NCO described the Soviet POWs as "dull, animal-like and ragged—and yet often treacherous,"[164] and another maintained that the Russians were "no longer human beings, but wild hordes and beasts, who have been bred by Bolshevism during the last 20 years." Therefore, he explained: "One may not allow oneself to feel any compassion for these people, because they are all very cowardly and perfidious."[165] Indeed, as early as 25 July Corporal Aloys Nackas maintained that the "enemy is not made of real soldiers, they are guerrillas and killers." But he was sure that "we at the front will finish off these Bolshevik hordes, for which the whole of Europe will thank us."[166] Karl Fuchs described on 3

August the obvious inferiority of Russian POWS: "Hardly ever do you see the face of a person who seems rational and intelligent. They all look emaciated and the wild, half-crazy look in their eyes makes them look like imbeciles." He simply could not understand how "these scoundrels, led by Jews and criminals, wanted to imprint their stamp on Europe."[167] But he had no doubt that the "war against these sub-human beings" was nearly over, and found it "almost insulting when you consider that drunken Russian criminals have been set loose against us. They are scoundrels, the scum of the earth! Naturally," he was happy to point out, "they are not a match for us German soldiers."[168] On 22 September Fuchs asked his wife: "Can you imagine that human beings grow up like animals? That seems to be the case here. . . . I suppose its just impossible to ask a Russian to think of something beautiful and noble," he concluded.[169] In another letter he described the "distorted, grimacing faces" of the Russians, "driven by a political insanity," and the rage which they awakened in him. "In my opinion, these Bolsheviks are murderers of all culture!"[170] Fuchs was quite sure he had come face to face with reality: "We have seen the true face of Bolshevism," composed to his mind of "Communist scoundrels, Jews and criminals," and, he assured his mother, he and his comrades "will know how to deal with it in the future."[171] And yet, if the Russians were so inferior, and Bolshevism so despicable, some soldiers asked themselves why were there "so many martyrs for the Bolshevik cause" among them. The answer supplied to them by the Wehrmacht's propaganda, and cited in many letters, was simply that there was "something diabolical about them."[172] But then both the size of the country, and the sacrifices made by the Russians to defend it, produced an even greater determination to wipe them out completely. As Günter von Scheven wrote in September 1941: "Spatially there is no goal, the landscape extends ever farther . . . the enemy is countless, though hectacombs have been sacrificed. Apparently everything must be annihilated before the war will end."[173]

Anti-Semitic sentiments among the troops increased as conditions at the front worsened and as soldiers were no longer merely exposed to racist propaganda but also observed and in some cases participated in mass murders of Jews. Whereas concerning the Russians soldiers occasionally expressed pity, the fate of the Jews only enhanced the feeling that this was a "race" which indeed deserved total annihilation, particularly as it might otherwise take revenge on the Germans for its destruction. But from the very first weeks of "Barbarossa" many soldiers' letters revealed the impact of years of anti-Semitic indoctrination and deeply rooted prejudices. Lance-Corporal Paul Lenz maintained early on in the campaign: "Only a Jew can be a Bolshevik, for this blood-sucker there can be nothing nicer than to be a Bolshevik. . . . Wherever one spits one finds a Jew. . . . As far as I know . . . not one single Jew has worked in the work-ers' paradise, everyone, even the smallest blood-sucker, has a post where he naturally enjoys great privileges."[174] In early August 1941 Lance-Corporal Herbert Nebenstreit wrote of his impression of Russia: "Only in Poland have I seen so much filth, mire, and rabble, especially Jews. I think that even there it was not half as bad as here."[175] Private Reinhold Mahnke furnished a detailed description of Bolshevik-Jewish atrocities against the Lithuanians. Not only did they eject them from their houses and then burn them down, they also "cut off their feet and hands, tore out their tongues. . . . They even nailed men and children to walls. Had these criminals come to our country," Mahnke now realized, "they would have torn us to pieces and mangled us, that's clear. But the Lithuanians have taken revenge," he concluded, referring to the anti-Jewish pogroms conducted by the local population with the encouragement of the *Einsatzgruppen* and under the observing eye of the Wehrmacht.[176] Lance-Corporal Heinrich Sachs similarly noted "how the Jewish question was solved with impressive thoroughness under the en-thusiastic applause of the local population." He then went on to quote Hitler's speech before the Reichstag threatening the Jews with destruction if they caused a war against Germany, and added that the "Jew should have known that the Führer was serious and must now bear the appropriate consequences."[177] Captain Hans Kon-druss, writing from Lvov (Lemberg) in mid-July had discovered ample evidence to show that "here clearly a whole people has sys-tematically been reared into subhumanity. This is clearly the most Satanic educational plan of all times, which only Jewish sadism could have constructed and carried through." The fact that the municipal library contained the Talmud, and that among the massacred civil-ians there were allegedly no Jews, was to his mind "indicative of the real originators." He too was satisfied to note that the "wrath of the people has however been turned upon this people of criminals." Indeed, he asserted: "It will be necessary radically to scorch out this boil of plague, because these 'animals' will always constitute a danger." The Jews had turned the population away "from everything which to us human beings has been eternally holy," for their goal was "the brutalization [*Vertierung*] of a whole people, in order to make use of it as an instrument in the war for Judas' world domination."[178] Lance-Corporal Paul Rubelt agreed that the "Jews were for the most part the evil doers" in the Lvov massacres, and noted that now the "culprits are shot."[179] Indeed, Corporal K. Suffner, who maintained that the "Bolsheviks and the Jews have murdered 12,000 Germans and Ukrainians in a beastly manner," reported that "the surviving Ukrainians arrested 2000 Jews and exercised frightful revenge." He concluded: "We swear that this plague will be eradicated root and branch."[180] Lance-Corporal Hans Fleischauer expressed similar sentiments: "The Jew is a real master in murdering, burning and massacring. . . . These bandits deserve the worst and toughest punishment conceivable." The consequences he drew from his experience with Jewish atrocities were far from untypical: "We all cannot be thankful enough to our Führer, who had protected us from such brutalities, and only for that we must follow him through thick and thin, wherever that might be."[181] Private von Kaull believed that "international Jewry," already in control of the capitalist world, had taken "as a counter-weight this proletarian insanity" as well: "Now these two powers of destruction have been sent to the field, now they are incited against Europe, against the heart of the West, in order to destroy Germany." He was impressed with the scale and significance of the conflict: "Such a huge battle has never before taken place on earth. It is the greatest battle of the spirits ever experienced by humanity, it is waged for the existence or downfall of Western man and the highest values which a people consciously carries on its shield." Consequently: "We must give our all to withstand this battle."[182] Private Gregor Lisch asked his family in the rear to "be happy that the Bolsheviks and the Jews had not come to us," for "the Jews have destroyed these poor people."[183] And Private Fallnbigl, while stressing that "we should be happy that we have not had this scourge of humanity in our own country," was convinced that "the German world would not be prepared for such heinous deeds even after years of preparation."[184]

As the war dragged on, soldiers progressively embittered by the endless fighting readily accepted the propagandistic line that the Jews were to blame. As one lance-corporal exclaimed in April 1942: "These swine of human creatures. They have clearly brought us this outrage of a war."[185] Typical of the inversion process common among the troops and the sense that the murderous treatment of the Jews merely confirmed their inhumanity was the following letter sent in July 1942:

> About events in the East concerning the Jews one could write a book. But it would be a waste of paper. You can be sure that they come to the right place, where they will no longer oppress any peoples.[186]

The frustration caused by partisan activities also contributed to anti-Semitic sentiments among the troops. It was the Wehrmacht's policy to execute large numbers of civilians in retaliation for any attack on military personnel, and the Jews were clearly the most convenient target, especially as the local population itself was often also strongly anti-Semitic.[187] The soldiers were quick to draw the conclusion that not only did the Jews constitute the main support of "Bolshevism" in Russia and had been about to overtake Germany as well, but that they were also directly responsible for the growing number of "terroristic" guerrilla attacks. One NCO wrote home in July 1942 that

> the great task given us in the struggle against Bolshevism lies in the destruction of eternal Jewry. Once one sees what the Jew has done in Russia, one can well understand why the Führer began the struggle against Jewry. What sorrows would have come to our homeland, had this beast of a man had the upper hand? . . . Recently a comrade of ours was murdered in the night. He was stabbed in the back. That can only have been the Jew, who stands behind these crimes. The revenge taken for that act brought indeed a nice success. The population itself hates the Jews as never before. It realizes now, that he is guilty of everything.[188]

It is interesting to note that the encounter with real Jews seemed to confirm even the most pornographic and malicious anti-Semitic propaganda produced in the Third Reich. Thus while it is true that initially it was easier to create hatred and fear of an abstract enemy, once this image had been internalized soldiers applied it to real living human beings, apparently believing that they actually resembled the caricatures of "the Jew" in Nazi newspapers. As one corporal wrote, although in the course of this war a little more light will have been cast on the Jewish question even for the most pigheaded philistine [*Spieser*], it is nevertheless still of the utmost importance that this question be further put in the necessary light, and here the *"Stürmer"* has, thank God, still remained true to its old positions. Just as the Eastern Jew now reveals himself in all his brutality, so have all this vicious lot, no matter whether in the West or in the East.[189]

Indeed, Streicher's yellow sheet seems to have enjoyed a great deal of popularity at the front. One NCO reported in August 1942 that

> I have received the *"Stürmer"* now for the third time. It makes me happy with all my heart. . . . You could not have made me happier. . . . I recognized the Jewish poison in our people long ago; how far it might have gone with us, this we see only now in this campaign. What the Jewish-regime has done in Russia, we see every day, and even the last doubters are cured here in view of the facts. We must and we will liberate the world from this plague, this is why the German soldier protects the Eastern Front, and we shall not return before we have uprooted all evil and destroyed the center of the Jewish-Bolshevik "world-do-gooders."[190]

This reaction to reality as a confirmation of propaganda was evident concerning other aspects of the Soviet Union as well. Thus early on in the campaign Private W. Lämmert wrote that "if in the past I thought that our propaganda had in this respect [conditions in Russia] somewhat exaggerated, today I can say that it had rather embellished conditions, for reality here is still far worse."[191] Lieutenant Lorenz Wächter also noted that what he had seen was "much, much worse than the German newspapers can write."[192] And Sergeant Hans Schimanowski, afraid that people at home might not have been give an accurate picture of reality in the East, recommended that they come to sample it. "I would never have believed that anything like this exists in the world," he admitted, "[b]ut we Germans are fighting for a just cause, and therefore victory will be ours."[193]

Retrospectively, at a distance of thirty years, and particularly when questioned by police officials, soldiers sometimes described events differently than in their letters from the front. The surviving members of a company which belonged to the elite 1st Mountain Division were interrogated in 1971 regarding the massacre of 317 civilians of all ages and both sexes in the Greek village of Komeno on 16 August 1943. Ostensibly in retaliation for a partisan attack, the officer in charge is said to have ordered the troops to "shoot everyone and leave nothing standing." The soldiers carried out the order, but as one man reported to his interrogators, "there was much discussion of this action" within the company. "Few thought it right," he went on to say. "I myself was so sickened by the atrocities [*Grausamkeiten*] that it took me weeks to recover my peace of mind [*seelisches Gleichgewicht*]." Another soldier noted: "Most of the comrades were very depressed. Almost none agreed with the action." Indeed, he asserted: "With some exceptions, all of them had crises of conscience [*Gewissenskonflikte*]." One NCO allegedly warned his commanding officer that this would be "the last time I take part in something like that. That was a disgrace [*Schweinerei*] which had nothing to do with fighting a war." Men are said to have asked comrades armed with automatic weapons to shoot in their stead, because they could not aim their rifles at the inhabitants. Yet another of those questioned reported that

> most of the soldiers did not agree with this action. . . . Many said openly that it was nothing but a disgrace [*Schweinerei*] to shoot unarmed civilians. Others, rather fewer, took the view that they were all a potential enemy so long as they supported the partisans against us soldiers. The argument was so heated that I might almost speak of a mutiny [*Meuterei*].[194]

Such testimonies seem to indicate that there was much more criticism of atrocities committed by the Wehrmacht among the soldiers than the official documents and, for that matter, private correspondence betray. As far as the troops' letters are concerned, this is not surprising, for negative comments about the regime and the implementation of its policies could and did lead to prosecution and heavy punishment. The fact that official documents also rarely mention such outspoken criticism may mean that officers preferred not to remark on negative attitudes among the troops in reports to their superiors. But as the court martial records of combat divisions also rarely refer to soldiers charged with criticizing reprisals of this sort, one may assume that this was far from a common phenomenon. It may also mean that as long as one carried out orders, even if reluctantly, one had no reason to fear disciplinary action for expressing one's personal disgust. Indeed, the most important aspect of this incident is that the men acted precisely as they had been ordered and massacred over half the population of the village, in spite of the fact that many of them were allegedly aware of the criminal nature of the operation. Moreover, as this reprisal action was carried out in 1943 by a unit which had previously served on the Eastern Front, where such massacres were quite common, it is unlikely that it was the troops' first experience with Wehrmacht atrocities. This would raise the question why were they so critical this time around. Perhaps the fact that on this occasion the action was against a Greek, rather than a Russian village, may have made them feel particularly uneasy. We have seen that Nazi propaganda was most extreme regarding the Russian/Jewish/Bolshevik enemy; it certainly did not put a particular stress on the Greeks. Thus there was a much less distorted, dehumanized image of the Greek inhabitants. More important, perhaps, is the fact that we depend here on testimonies given in the course of a police inquiry thirty years after the event. It is more than likely that these circumstances had a major distorting effect on the men's recollection, both because of fear of prosecution, and because the growing awareness over the years that they had actually taken part in a massacre may have caused them to exaggerate their objections to it at the time. As one man who had been a junior officer in Poland and Russia said under less delicate circumstances forty years after the event:

> Well, of course, what they [the Nazis, rather than "us," the Wehrmacht] did to the Jews was revolting. But we were told over and over again that it was a necessary evil. . . . No, I must admit, at the time I had no idea we had fallen into the hands of criminals. I didn't realize that until much later, after it was all over.[195]

We can thus conclude that while this and similar incidents were accompanied by a certain amount of verbal resistance, criticism was probably far more limited than former participants preferred to remember; that as long as one carried out orders, one was not punished for expressing reservations; and that criticism was more common when non-Russians were involved. Reprisals were carried out throughout Europe, but on an incomparably greater scale in the Soviet Union; and, as Soviet citizens were portrayed by Wehrmacht propaganda and seen by many soldiers as *Untermenschen*, not only were such actions executed without any resistance from the troops, they also gave cause to very little criticism. In fact, as the soldiers' letters demonstrate, they were often viewed with a great deal of approval. Hence the case of Komeno may be described as the Greek exception which proved the Russian rule.

Faith in the Führer was from the very start a major component in the distortion of the troops' perception on the Eastern Front. In mid-October 1941 one NCO expressed this feeling succinctly when he wrote that "for us the Führer's words are gospel."[196] Private von Kaull exclaimed in a letter to his brother:

> The Führer has grown into the greatest figure of the century, in his hand lies the destiny of the world and of culturally-perceptive humanity. May his pure sword strike down the Satanic monster. Yes, the blows are still hard, but the horror will be forced into the shadows through the inexorable Need, through the command which derives from our National Socialist idea. This [battle] is for a new ideology, a new belief, a new life! I am glad that I can participate, even if as a tiny cog, in this war of light against darkness.[197]

As the fortunes of the *Ostheer* rapidly deteriorated, the troops' "belief" in Hitler did not falter, but rather increased in direct proportion to the hopelessness of the situation. While at a time of great victories praise of the Führer was accompanied by a confidence in the Wehrmacht's own invincibility, the growing sense of the army's inability to overcome the military crisis created a need to rely on an irrational faith in the only man who was perceived as Germany's destiny, for better or for worse. Like all gods, Hitler's ability to mold the course of history was derived from the faith of his followers. Thus belief became both a personal, psychological need among the troops, and a weapon which would strengthen the Führer and enable him to wring from history the now increasingly mythical *Endsieg*. The battle of Stalingrad was a good indication of this process. The Field Post Inspection Office (FPP) of 4 Panzer Army examined over 11,237 letters sent to and from Stalingrad between 20 December 1942 and 16 January 1943, and noted that almost until the very end the soldiers kept expressing their loyalty to and belief in Hitler. In early January 1943 one corporal wrote: "We all have the staunchest hope that the Führer will not abandon us and will surely find a way out, as he has always done before." Another soldier maintained that "we stand in complete trust of the Führer, unshakable until the final victory." As late as 16 January one of the soldiers claimed that the men were certain the siege would soon be broken. "This is why we hold on as strong as iron," he explained, for the salvation was "as certain as 2 x 2 = 4." Another soldier forcefully described the profound crisis of belief in which he found himself even when reality provided him with ample evidence that all was lost and that Hitler had abandoned the men of Stalingrad:

> The Führer has promised to get us out of here. This has been read to us, and we all firmly believed it. I still believe it today, because I simply must believe in something. If it isn't true, what is there left for me to believe in?... Let me go on believing, dear Grete; all my life—or eight years of it, at least—I have believed in the Führer and taken him at his word.... If what we were promised is not true, then Germany will be lost, for no other promises can be kept after that.[198]

Of course, by this stage the intense hunger, cold, and rapidly deteriorating military situation were having an effect. Soldiers now wrote: "One becomes apathetic about everything and thinks only about eating." Yet even among those who realized that they were doomed, some felt certain that their sacrifice would not be in vain and had a higher significance which transcended individual fate. Rather than admit that they had been abandoned by Hitler, these men accepted even now the propagandistic line that the military disaster whose victims they had become was a necessary step toward the *Endsieg*. Thus one NCO wrote home that

> the world outside should also know what German soldiers have taken upon themselves and have been able to bear here. This heroism and this perseverance are unprecedented and will be rewarded by a great victory.

The refusal to accept the futility of the battle thus led many soldiers to grasp at any ideological interpretation which presented the debacle of Stalingrad as a victory of "world-historical" significance. As one major wrote to his wife:

> The pitiless battle goes on, the Lord helps the brave! Whatever providence may have decided, we ask only . . . to have the strength

to hold on! One will say of us one day, the German army fought in Stalingrad as no soldiers in the world had ever fought before. It is the mothers' task to pass this spirit on to our children.[199]

While the men inside Stalingrad were naturally haunted by fear and doubts as their end approached, many of those fortunate enough not to be swallowed into the cauldron seem to have been taken in by the propagandistic reinterpretation of the disaster as a powerful rallying point for even greater sacrifices. Instead of seeing it for what it was, the beginning of the end and a clear sign of Hitler's senseless strategy, one lieutenant wrote from the front in early February 1943 that

> this struggle for life, this looking death straight in the face, is quite inhumanly heroic! There in Stalingrad now surges a sea of the best German blood. . . . I believe that never before has National Socialist Germany been as earnest as now. . . . Here it is no longer a matter of the individual, here it depends on the whole. Only as long as we are aware of that, will we be able to achieve victory![200]

Soldiers did not merely accept the propagandistic claim that great sacrifices justified even greater bloodshed, but were also increasingly persuaded by the social-darwinist argument that both the winners and the losers deserve their fate. The only way Germany could prove that it had fought all along for a just cause was by winning, no matter by what means and at what price. As another officer wrote, the war

> compels us to the *greatest* exertion of *all* our powers in order to pass through the moments in which one feels that it is *beyond* our powers. . . . If only because the continuation of this war can certainly *not* bring us greater misery and stress, than if we were to put down our arms. I would not like to think what would happen then! . . . *Now* is the test: if we hold on now, then we shall have a future. If we do *not* hold on—then we do not deserve any future, then we do not deserve any pity.[201]

Thus the soldiers progressively retreated to an unreal, mystical, nihilistic world whose features had little to do with their actual experiences at the front, but were taken from propaganda tracts written by men who had mostly never been in a battle, or had long forgotten its realities and preferred to glorify its memory. The remarkable paradox was that the closer one was to the front, the more earnestly one tended to take the heroic verbiage supplied from the rear. Not a few historians, perhaps because most of them have also not experienced battle at close quarters, believe that combat soldiers are the first to see through the glorification of war commonly associated with "soldiers of the pen"; there is good reason to doubt whether this is the case in any army, especially regarding national conscript forces.[202] But the case of the Wehrmacht was certainly different, for the troops at the front were the firmest of Hitler's followers, and the least cynical about his ideology.

Increasingly during the last two years of the war, the troops at the front came to see themselves as the missionaries of the entire German nation, indeed of Western civilization as a whole. Rational evaluation and clear perception of events were replaced by intense terror from and rage against a faceless, monstrous enemy, which in turn only enhanced the men's desperate clinging to their faith in Hitler's ability to avert the apocalypse and lead the Reich to the *Endsieg* over the forces of evil. It was at this period, just as Germany was accelerating even further the implementation of its genocidal policies, that the view of the Wehrmacht as the protector of humanity gained increasing force. Paradoxically, the soldiers' awareness of the regime's criminal actions (at least at the front) made them fight with even greater determination for its survival by intensifying their fear of the consequences of defeat. Note the following letter by a Wehrmacht captain written in mid-February 1943:

> May God allow the German people to find now the peace of mind and strength which would make it into the instrument needed by the Führer to protect the West from ruin, for what the Asiatic hordes will not destroy, will be annihilated by Jewish hatred and revenge. The belief at the front is unshakable, and we all hope that, as Göring has said, with the rising sun the fortunes of war will again return to our side.[203]

This was, indeed, the core German troops' ideological motivation, a combination of prejudices and phobias which made them so much into Hitler's soldiers. God was with the Führer, and the German people were God's instrument, whose goal was to save the West from Asiatic barbarism and Jewish revenge. The danger was great; but as long as belief in Hitler remained unshakable, victory was certain to come. Ironically, even men who claimed that the "time of fanaticism and intolerance of other views is over," and that "if we want to win the war, we must become more rational" concluded that all this was necessary "so that we will not be delivered to the revenge of the Jews."[204] By this stage of the war, very few soldiers could remain immune to Nazi terminology and modes of thinking, even if they saw themselves as opponents of the regime. The most sceptical were found willing to sacrifice themselves for what ultimately meant the Hitler regime, and refused to consider the possibility of giving up or rebelling against the insane orders of their superiors. Friedrich Böhringer, a twenty-year-old student of philosophy, readily admitted on 22 March 1944 that it was "after all difficult to stand in front of a dark abyss and to be told: leap in and bury yourself in it! This is your destination, nothing else!" He asked: "Has there ever been a young generation with such uncertainty and blackness, with such a bleak future as ours? Yet," he insisted, "it is still necessary once again to overcome the horror of natural man and to let the other side of us be expressed, that which knows the eternal laws." Böhringer would no longer be "consoled with the daily slogans of the politicians. . . . This sham-world is not worth dying and fighting for!" But: "For Germany? Of course—for the hidden, eternal, Germany! Is it still necessary to make big speeches about this?"[205] And the twenty-one-year old Franz Rainer wrote from Italy on 26 August 1944:

> The Lord must still see and once more help us out of this predicament. Are we really the wicked who have been condemned to destruction? . . . Will the others still succeed to triumph over us, we who had ventured forward too far and already believed ourselves to be like God?[206]

The failed *Putsch* attempt served as a particularly powerful indication of the reserves of "belief" in the Führer still remaining among the troops of the Wehrmacht only months before the Third Reich finally collapsed. Indeed, Hitler's "salvation" even further enhanced his divine aura and appeared conclusively to prove God's approval of the Führer and his actions, as he himself so often asserted. Conversely, the generals' plot also supplied the soldiers at the front with an explanation for the endless defeats they had experienced in spite of their tremendous efforts and Germany's allegedly God-sent leadership. Now all failures and mistakes could be attributed to the assassins who, in league with the Jews, Bolshevism, and Plutocracy had formed a fifth legion in the rear and had almost stabbed Germany once more in the back. The conspirators knew well enough that such would be the reaction of the troops if they failed;[207] yet it is still remarkable to what extent the Putsch of July 1944 rallied the soldiers around Hitler and motivated them to go on fighting in total disregard of objective military reality. One corporal raged against "this vile booby-trap against the Führer" in a letter written the next day. "Thank God," he went on to say, "once again he survived it." Reassuring the rear, he then concluded: "Among us here there is a general feeling of indignation over this crime."[208] For the man at the front, any attempt to dispense with Hitler was a crime; the regime's own crimes were moral actions derived from a historical and existential necessity. One lieutenant wrote that it was "morally depressing" and "criminal . . . to want to do away with such a deserving people's leader."[209] Hitler's salvation, however, was clearly an act of God. As one corporal put it: "Providence has protected our Führer from everything." Indeed, this man could well imagine "[h]ow the people will rejoice and with what happiness it will hear the news that the beloved Führer still lives . . . because he is the one who leads us to the *Endsieg*."[210] Another lieutenant wrote two days after the *Putsch* attempt that his colonel had "had tears in his eyes" upon receiving the news, and went on bitterly to rail against "the swine who had thrown the Führer to the devil," opposing them to his own "front and troops generals . . . who believe unshakably in the Führer." Here was the same symptom of greater confidence in Hitler the closer one came to the front, not unlike the parallel phenomenon of greater religious observance among troops immediately on the front-line.[211] The need of soldiers under constant danger of death for some kind of spiritual support, provided in the Wehrmacht first and foremost by a quasi-religious belief in Hitler, was thus powerfully demonstrated in this period of profound military and psychological crisis. Thus the letter cited above concluded with the revealing statement that the "wretched lot" of conspirators had failed to "take away from us our belief in victory," and that Hitler's survival was "the great providence, which can only strengthen our belief even further."[212] After all the great sacrifices, it was felt, the *Putschists* had almost snatched victory away from Germany at the last moment, as had indeed happened in the Great War. One lieutenant exclaimed: "These bandits tried to destroy that for which millions of people have already given up their lives." Yet their failure gave him "a nice feeling," because now he knew "that a November 1918 will not be repeated."[213] Another combat soldier wrote:

> I and all members of the company were struck dumb by the announcement of this revolting infamy. Praise to God that providence has preserved our Führer for the salvation of Europe, and it is now our holiest duty to bind ourselves to him even more strongly, so as to make good again what the few criminals, probably paid by the enemy, have committed without any consideration for the people as a whole. . . . They all belong to the public gallows.[214]

Forty years later, a man who had served in the war as a combat soldier, and whose father had been a Wehrmacht major-general, illustrated much of the grotesque insanity of the period in the following personal account:

> It came as a terrible shock to me. My whole world collapsed. I was a young first lieutenant at the time, serving on the Eastern Front. When my commanding officer informed me that my father had been executed for plotting against the Führer, I thought he was out of his mind. My father—a dedicated National Socialist and a model officer—it was unbelievable. Why should he have gone over to the opposition? We owed the Führer everything![215]

These were anything but isolated outbursts. In August 1944 a report by 3 Panzer Army's FPP, based on the inspection of 44,948 letters, stressed:

> The high number of expressions of joy at the Führer's deliverance, presented as a real stroke of good fortune for the German people, prove not only the love and loyalty of the soldiers to the Führer, but also reveal . . . the soldiers' strong determination to fight and win in his sense.

This was in spite of the fact that a quarter of the letters contained complaints regarding the general situation, the men's direct superiors and the higher leadership, as well as on lesser issues such as provisions, friction within the units, and postal delays. Thus although a significant rise in criticism was observed, these comments entirely excluded Hitler and everything that he stood for. Indeed, remarks considered by the censor as constituting serious breaches of discipline, sabotage, or subversion were found in only 50 letters, that is, a mere 0.1 percent of the overall number examined; a particularly impressive achievement taking into account the terrible mauling 3 Panzer Army had only recently experienced in the Soviet summer offensive of June 1944.[216]

Throughout the last months of the war, and in some cases merely a few weeks before the collapse of the Reich, many soldiers refused to face up to reality, and continued expressing unflinching trust in Hitler, no matter how obviously false his promises of victory had proved to be. In August 1944 one corporal insisted:

> When the Führer says that we have the means and the weapons to push the enemy back once more from our borders, and that we will ultimately wring victory from him, then I know very well that an unbending trust and a strong and uncompromising belief in the Führer are essential to overcome this momentarily difficult period. . . . Belief gives the strength to bear all hard and difficult sorrow. . . . My belief in the Führer and in victory is unshakable.[217]

Escape from reality was of course a constant feature of the troops' reaction to the war. Even in August 1941 Siegbert Stehmann, a twenty-nine-year old theologian, expressed that hodgepodge of romantic philosophy, religion, and fascist ideology so typical of his generation in a letter written following a battle in which 80 soldiers and all officers of his company were killed: "To us lonely men one thing has been revealed in the hopelessness: that reality is nothing, miracle [*das Wunder*] is however everything. This keeps us going. No man can help us, only God alone."[218] This nihilist romanticism greatly increased toward the end of the war even among the most disillusioned of Hitler's soldiers, and made them ready to sacrifice themselves for what they still believed would be a better future. Indeed, by September 1944 Stehmann was writing:

> How the horror of these times has influenced us, that we accept with indifference the all-present terror which we would have never been able to imagine! The German people has almost surpassed the legendary capacity for suffering of Russia. Perhaps this is the greatness of its hour. . . .
>
> Germany, the eternal Job of world history, sits everywhere on

the ruins of its still, beloved world, and waits passionately for the releasing word of God, who can heal the broken. . . . The material worries weigh lightly, when one contemplates our *Volk*, when one thinks of the approaching fate of our *Volk*, whose country has already been penetrated by the enemy. . . . A thousand-year Reich is going to the grave. . . . God will help us. . . . No one in the world is more blessed than our *Volk*, which even today sends its roots deep into the earth.[219]

Reinhard Becker-Glauch, an art historian aged twenty-eight, wrote on 27 September 1944, shortly before his death:

The great, sweeping slogans of this war have grown dumb, and over all it is now a question of mere survival. Yet even with such constricting goals war has its greatness and its elevating happiness, more than any other, because it drives us all consciously to the limit of things; the sham-values sink behind and what really occupies the heart, you and the homeland, alone remain strong. These are more than platitudes. Hour by hour we experience the blazing love to you and to the Fatherland as the driving realities.[220]

Klaus-Degenhardt Schmidt, a torpedo boat commander aged twenty-six, similarly asserted in December 1944: "My goal in this war is the formation of the *Volk*, the end of all previous history. . . . For me the *Volk* is a temporal law for which I have stepped forth in affirmation of divine orders. I believe in its holy destination and goals, in its reality as a decree of providence. It fights for its existence against a world." Indeed, Schmidt did not perceive of an end to the struggle: "After the weapons are laid down [the *Volk*] will have to undergo a spiritual battle to the end. We need sacrifices and assistance. This is the birth of the hidden and the visible Germany." Thus death was birth, defeat was victory, suffering was purification: "Each year of misery and war was a school, whose sense is already evident in spite of all sacrifices." What that sense was, Schmidt did not reveal. But in his last letter he reflected "on everything beautiful, eternal, light, renewable in us and around us," and found that "this time enables us to accomplish more with joy and goodness of heart than the more comfortable years of the past." Thus he went down fighting for Hitler and imagining himself to be dying for a better world.[221]

Waiting for the promised *Wunderwaffen* increasingly combined

with terror from the consequences of defeat, vividly portrayed by
the Wehrmacht's propaganda and effectively motivating the troops
to go on fighting out of fear of the enemy precisely when the number
of Hitler's faithful finally began to diminish. In September 1944 one
lieutenant maintained:

> Nothing may exist which could make us weak. Any German defeat
> would spell a total... destruction of all Germans. ... We are the
> last bastion, with us stands and falls everything which has been
> created by German blood over the centuries.[222]

Still many soldiers refused to accept that the end was near. Rolf
Hoffmann, aged twenty, wrote on 4 February 1945:

> Six long years we have held our own against a world of enemies.
> ... Do we deserve finally to break down and be destroyed? We
> want to have confidence in God, that He will not abandon the
> German people and will give back to it at the end of this mighty
> struggle its right to live on this earth. Therefore we must endure,
> until a better future is assigned to us.[223]

Wilhelm Heidtmann, a thirty-year-old candidate of theology, wrote
in August 1944: "God grant me always to do that which is needed
for the protection of the homeland, and to be certain with all my
heart!" By September he asserted that "God's kingdom cannot be
conquered with human weapons," from which he concluded that
"even the Western powers will not be able to prevent the fulfilment
of that which Christ had said of his resurrection," namely, the
rebirth of the German Reich. And as late as March 1945 he reported
that "some of the [British] enemy waved to us: 'comrades!'—but
German paratroopers do not cross over to the other side."[224] Dr.
Clemens August Hoberg, aged thirty-three, wrote on 24 February
1945:

> We held our positions in Pomerania to the last moment. ... The
> general situation is certainly progressively worsening, and it is not
> difficult to foresee that the events of this year will bring with them
> the climax and the end. For there is nothing left but to defend
> ourselves to the last. Any capitulation would only spell our certain
> end. ... So long as the battle goes on, we still always have innu-
> merable incalculable possibilities.[225]

Fear and distrust of the enemy also meant that the troops rejected the propaganda directed at them by their opponents, as the higher military and civilian authorities of the Reich acknowledged with satisfaction and the Allies were forced to admit.[226] Indeed, one Panzer captain believed that only "limitless hatred and the ultimate sacrifice" were the proper answer to the enemy's propaganda.[227] Many soldiers asserted once more that the enemy was even worse than his portrayal by their own propaganda. While one corporal was simply filled with horror at the thought that "the black and yellow races will destroy and devour Europe,"[228] a lieutenant pointed out that "what the newspapers write is only a watered-down version of what the Bolsheviks actually do wherever they come."[229] By the last weeks of the war it was probably terror more than anything else that sustained the troops at the front.[230] In mid-March 1945 one soldier wrote that "we still hope that our culture will be spared from the Mongol stream."[231] But there were still those who hoped to continue the good fight, either in this war, or if it were truly and finally lost, which they were still reluctant to admit, in the next.[232] This was certainly the opinion of a soldier writing from a military hospital in late March:

> Even if the war turns out unfavourably for us, which, as I have said, is still far from clear, then I am not one to hide my head in the sand. . . . As the alliance between the Western powers and Stalin is merely instrumental, it seems to me that there will still be at least one more confrontation between these two fundamentally opposing parties. For this, as well as for our own battle against Bolshevism, the German people will in any event still have to bring more sacrifices. In any case we have still not lost the war.[233]

Nor was this soldier entirely wrong. Repeating what the Nazi regime maintained at the time and, along with many others, anticipating the Cold War, his letter also contained the essence of the argument made much more recently by revisionist (and from a different angle also by left-wing) historians, namely, that "Hitler's war" had been waged first and foremost against communism and consequently, as some will have it, that the Third Reich had served as the main bulwark against "Bolshevism" and ought to be given its due for having "saved" Western civilization from Asiatic barbarism.[234] According to this view it was the failure of the Western

Allies to grasp the "true" nature of the war that made them insist on the destruction of the Reich, a mistaken policy which led to the occupation by the Red Army of Eastern Europe, instead of teaming up with the Wehrmacht to push back the Russians. Indeed, the belated realization of this crucial aspect of the war formed the basis for the resurrection of the Bundeswehr as the successor of the Wehrmacht in its central role of keeping communism out of the West. These are arguments which must be kept well in mind when one ponders the significance of German reunification, for the creation of a German superpower in Central Europe will fill the void created by the retreat of the Soviet Union and make possible the formation of a German dominated *Mitteleuropa*, a concept which can be traced back all the way to the original unification of Germany, if not indeed to the aspirations of German nationalists from the middle of the nineteenth century. A less distorted view of the past will thus make for a better understanding of what the future still holds for Europe as it moves to what some would like to call the postwar era, and others might feel may be a reversal of history to the instability of the *entre deux guerres*. It is in this context that we should recall Hitler's own postscript to his testament, dictated on 2 April 1945. Germany's defeat, he said, would be a tragedy for Europe as well as for the German people:

> With the defeat of the Reich . . . there will remain in the world only two Great Powers capable of confronting each other—the United States and Soviet Russia. The laws of both history and geography will compel these two powers to a trial of strength, either military or in the field of economics and ideology. These same laws make it inevitable that both Powers should become enemies of Europe. And it is equally certain that both powers will sooner or later find it desirable to seek the support of the sole surviving great nation in Europe, the German people.[235]

Alan Bullock considered these words prophetic when he cited them in his book some thirty years ago. How much more prophetic, and menacing, they may seem to some readers today.

To conclude, we have seen that throughout the war, but especially between autumn 1941 and almost until the "capitulation"—a period characterized by a growing sense of crisis accompanied by an increasingly mystical hope for salvation—the troops' view of reality was composed of two interrelated elements: a progressive dehumanization of the enemy and a parallel deification of the Führer. In the case of Hitler, as his opponents realized, only death could undo his god-like image and free his followers from his hold; in the case of the enemy, under the circumstances of postwar occupation and political and military impotence his demonic image had to be suppressed and repressed. Moreover, following the division of Germany, the dual character of the original enemy was divided among the two German republics: capitalism remained the enemy of the GDR, communism that of the FRG. This allotment of enemies was useful in facing up to postwar realities, not only because it was encouraged by the respective occupiers and became an essential part of the two Germanys' legal systems, but also because it did not necessitate a total break with former beliefs. Only hatred and fear of "the Jew" was officially done away with in both states, and has since had to find an outlet either in philo-Semitism, indifference, and denial, or to appear in more or less concealed forms among the more extreme right- and left-wing fringe.[236] It remains to be hoped that the powerful prejudices regarding the "other" which still existed under the surface in both republics will not come to play a larger role in the newly reunified and more independent Germany which has now emerged in the heart of Europe.

Conclusion

The German people did not go to war in 1939 with the same "*Hurra-Patriotismus*" of August 1914. Indicatively, the mood of the nation during the Sudetenland crisis in autumn 1938 was not essentially different from that of Britain and France.[1] Contemporary reports noted that there was "no enthusiasm whatsoever for entanglement in war," that "[m]orale [was] widely depressed," and that there was "overall a 'general war-psychosis.'" As the crisis reached its climax it was reported: "Everywhere there prevailed great tension and anxiety, and everywhere the wish was heard: anything but war... expressed with particular vigour by veteran World War front soldiers." Yet once the Munich accords were signed, the public mood rapidly changed into one of admiration for the Führer's "political skill." People had taken Hitler's "peace campaign" of the mid-1930s quite seriously, and had come to believe that he wanted to avoid war just as much as they did. By 1939 the public was so certain of the Führer's ability to extricate Germany from political crises without going to war that there was far less "war psychosis" than during the preceding year.[2]

When war did eventually break out, a widespread depression was noted by all observers of the German public. Bernt Engelmann recalls: "No crowds had gathered. We saw no trace of rejoicing, certainly none of the wild enthusiasm that Germans had shown... in 1914."[3] An American correspondent found the population apa-

179

thetic. A Gauleiter who was travelling throughout the country at the time remarked that there was "no happiness, no joy." Instead, he wrote, "everywhere I came an oppressive silence, not to say depression, prevailed. The whole German people seemed to be struck by a paralyzing horror, so that it was capable of neither expressions of approval nor of disgust." Another observer spoke of the "dull obedience of a mass educated to follow [its leaders] blindly and thoughtlessly, but also stupefied and confused by militant propaganda." It has been claimed that the majority of the German population manifested at the beginning of the war a mood of "unwilling loyalty."[4] One might say that the Germans accepted the outbreak of war with the same fatalism that characterized their behavior during the last desperate months leading to the "capitulation."

The great military triumphs in the first two years of the war dispelled much of the gloom. If in 1939 the Germans still believed that Hitler could always avert war at the very last moment, after the victory over France they were sure that he could defeat any enemy.[5] There were other reasons for contentment. During the 1930s the regime had done away with unemployment and was perceived by large sectors of the population as having brought Germany out of what had seemed until the *Machtergreifung* as an insoluble economic crisis. Prosperity and social order under Hitler's rule could be seen as a "return to normality" after the anarchy of the Weimar Republic's last years.[6] Not that the Nazis had created the idyllic *Volksgemeinschaft*; indeed, toward the outbreak of war there were renewed signs of unrest among the workers who wished to make use of the increasing labor shortage to demand higher pay.[7] Yet as long as one did not belong to the political or "racial" categories persecuted by the regime, was not subject to the "euthanasia" campaign, and did not engage in anti-Nazi activities, one was not badly off in Germany of the late 1930s.[8] Furthermore, when within less than a year Germany came to occupy most of Western and Central Europe, the economic prospects of ruling over the entire continent left little room for social unrest. National pride in Germany's military achievements combined with hopes for unprecedented prosperity from which everyone—at least every "Aryan"—would gain.[9] Successful empire-builders have always benefited from this kind of "social-imperialism."

The invasion of the Soviet Union caught not only Stalin, but also many Germans by surprise, and once more unleashed a wave of apprehension and anxiety. On the second day of "Barbarossa" the SD noted that "according to reports... from all parts of the Reich the announcement of the outbreak of war with Russia has caused great surprise among the population."[10] By July it was claimed that the "general mood [*Stimmung*] among wide sectors of the population has... undergone an increasing deterioration." Of course, stressed the SD, "by and large confidence in victory and trust in the leadership have remained," but "the people are depressed or embittered and incensed by the toughness of the struggle in the East, the criminal conduct of the war by the Red Army, by the noticeable casualties... and above all by the consequent difficulties of supply."[11] This was a telling analysis of the public mood, for it encapsulated the most important aspects of the war in the East—which from now on became *the* war—as perceived by the German population. Here was a mixture of irrational, but powerful terror from the Russians, and a very concrete worry about the material consequences of fighting a war on such a scale. Fear of "Asiatic Bolshevism," anchored in long established prejudices and fanned by Nazi propaganda, created the basis for a grotesquely distorted view of reality. Although it was known that Germany had attacked, the Soviet Union was seen as the aggressor; and while the public was promised to gain from the ruthless exploitation and enslavement of the Russians, the Red Army was believed to be the real criminal. Similarly, fears of shortages were accompanied by a widespread zeal for economic expansion which made the idea of a *Raubkrieg* in the East particularly popular among Germany's industrialists and business community.[12] The prospects of power and wealth were greater than ever before, but so were the risks. Hitler's previous triumphs had convinced soldiers and civilians alike that this war too would be won, but the enemy's determined resistance, and the terror from his dehumanized image, filled the Germans with great anxiety. The cause of that fear had to be wiped out by every means available. This was necessary because the enemy was evil; it was bound to succeed, because it was the Führer's wish; and it was also to be richly rewarded. In this sense the Germans were indeed fighting Hitler's "war of ideologies."

The war in the Soviet Union demanded increasingly total manpower mobilization. Even the German workers, who throughout the 1930s had resisted the myth of the *Volksgemeinschaft* and had striven to protect their economic, if not their political interests, were now drawn into the war and forced to become part of the no less mythical *Kampfgemeinschaft*. In the Wehrmacht the working class disappeared, only to re-emerge in 1945 after many years at the front as Hitler's soldiers and the representatives of the *Herrenrasse* in the vast territories occupied by Germany. At home some of them might have remained immune to the regime's propaganda,[13] but once in uniform they were sucked into the army's "melting pot" and forged into Hitler's instruments, becoming the executors of his policies, the conquerors of his empire. There were no class-oriented rebellions in the Wehrmacht, indeed, there were no mutinies at all. In the Ruhr industrial region workers might have grumbled against the regime, but in the ranks they numbered among those very same soldiers considered by all observers as Hitler's most loyal supporters. This transformation of Germany's workers into Hitler's soldiers was a measure of the regime's success in mobilizing the whole nation to fight its war of conquest and destruction. Naturally, the soldiers fought for many reasons; they fought for survival, for their comrades, for their families in the rear, and for Germany's victory and prosperity. But workers or not, they also fought against "Plutocracy," "Asiatic barbarism," and "Judeo-Bolshevism," and in defense of "German culture" and "Western civilization." In this sense they fought for Nazism and everything that it stood for.[14]

Perceived as a struggle of all-or-nothing, the war in the East called for complete spiritual commitment, absolute obedience, unremitting destruction of the enemy. As such the war in the East constituted not only the climax of the Nazi regime, but also the most important element of its postwar memory in Germany. Surviving its reality and living with its recollection necessitated a profound process of inversion. When the soldiers returned home, they brought with them the images and horrors of the war, the perverted morality which had formed its basis, and the distorted perception which had made living through it bearable. All of these were combined into Germany's collective memory of the war, for only thus could postwar society "come to terms" with its past. "Auschwitz" could be ascribed to a minority, numerous as well it might have been. Not so the war. Every family had sent at least one soldier to the front. Worse still, the young men who returned from the fighting became Germany's workers and bureaucrats, professors and techno-crats, bankers and politicians, judges and lawyers, writers and poets.[15] It was inconceivable that they had all taken part in a huge criminal undertaking. Thus the same psychological mechanism which had facilitated fighting a barbarous war was employed to facilitate living with its memory. The soldiers' combat experiences were portrayed as unique, the uniquely criminal characteristics of the war were "normalized" as mere by-products of the fighting. Cause and effect were reversed: barbarism was perceived as the outcome of the enemy's bitter resistance to occupation, not as its main trigger. The troops' sufferings were vividly remembered, their victims' were repressed. Nor was this view of the war limited to the soldiers. Civilians too perceived the strategic bombing raids and the occupation of the land by enemy armies as unique manifestations of the destructiveness of war; the industrial murder of millions of human beings was viewed as merely one more aspect of war's evil nature, not something unique to Nazi Germany's very specific war. And, because the bombing and occupation were experienced by many more Germans than the death camps, they remained implanted much more deeply in their memories.[16] Thus the genocide of Jews and Gypsies ceased to be directly related to the Germans, and be-came something executed not by them, but in their name. The culprits could not have anything to do with the resurrected German people of the postwar era. The war remained a deep, painful mem-ory, but it was a memory of one's own suffering, and it left no room for one's victims. If for Hitler the war had been a vehicle for winning over those Germans who had previously remained aloof from his regime, it served postwar German society to repress the memory of its crimes by lamenting its own fate.[17] The war had made the Wehrmacht into Hitler's army, the Germans into Hitler's peo-ple. Defeat converted them all into victims. If Austria was Hitler's first victim, Germany was his last. And victims cannot be called to account.

In 1981 the sixty-six-year old clerk Werner Paulsen said in an interview: "On 30 June [1941] I experienced the worst day in my life." He was one of many Germans whose memory was still dom-inated after forty years not by family matters, careers, or political events, but by the war; and, in that war, by very specific incidents which had marked them for the rest of their lives. This overwhelming memory of the war, and the extremely selective assortment of details it contains, is both a very intimate part of each individual's identity, and a major component of this entire German generation's collective consciousness. For the former NCO and platoon leader this was a Russian ambush: "All at once we were shot at from all directions. In front of us. Behind us. Bullets everywhere." Of Paulsen's platoon which had numbered fifty men, only four survived. Within an hour a total of 92 German soldiers were killed next to him: "one fell to the left, one fell to the right. It just went: putch, putch, putch. And there they all lay. It happened so quickly, one couldn't do a thing." The next day one of his comrades went to bury the dead: "And what he saw there was ghastly." The bodies were terribly mutilated. "Eyes gouged out. Genitals cut off." They had been fighting next to an SS unit, but Paulsen at first claimed not to know "whether the SS had already done this before." Later on in the interview he admitted that "on our side things were certainly not much different, then." His conclusion was common to many veterans: "Unfortunately it was really like that: what you do to me, I do to you."[18] Paulsen remembered this incident with great clarity; it was, after all, the worst day of his life, his closest encounter with death. But what he remembered was a bloody battle in which many of his comrades were killed and mutilated by the Russians. He conceded that the Germans had probably also committed atrocities, but here his memory failed him. Indeed, he may well have added this qualification due to his exposure to postwar media reports on Germany's criminal conduct in the war. He thus presumed that such events had taken place, but he could not remember taking part in them, or even observing them. The war was the single most important event in his life, and the most important element of that war was a massacre of German soldiers by Russians. This was *his* war, and that of most other survivors; this was also the collective memory of the war. As for the rest, it might have happened, one would do better not to deny it, but it formed no part of one's own experience and memory. In this sense, it had not *really* happened after all.

Thus the popular collective memory of the war is one of a terrible event in which many people suffered and died as victims of an apocalypse beyond their control. The impressive efforts of German scholars to document the involvement of numerous institutions and individuals in the very down-to-earth and practical aspects of the regime's criminal policies seem to have had little effect on this image of the war. This is the reason that every new study on the participation in murder of soldiers, or doctors, or lawyers, or any other profession or sector of society is invariably greeted with a sense of shock and astonishment. The facts are there for everyone to behold, but they are kept well apart from one's own experience and memory. Paulsen will admit that the Germans too may have committed atrocities; but he had no part in them, nor had any of his comrades. Indeed, this collective amnesia is so powerful that it infects even those very scholars who have done so much to cure it. Thus, for instance, an historian who had written the most substantial study of Nazi indoctrination in the Wehrmacht concludes by doubting its effect on the troops, and notes, furthermore, that he personally cannot recall having ever been exposed to political education during his own service in the Wehrmacht.[19] Another historian, who had done much to show the involvement of the lower ranks in the realization of the "final solution," states unambiguously that "[t]he mentality of the average *Landser* was characterized by soberness, [and a] rejection of the far-from-reality propaganda tirades," and that the troops "had little influence" on, and "could hardly find a way of avoiding the escalation of violence," which was in any case part of "the conduct of war by both sides."[20] Finally, yet another scholar, who had made a major contribution to our knowledge of Hitler's plans and preparations for the *Vernichtungskrieg* in the East, asks his (German) colleagues to identify with the German soldier, whose exploits he then describes with a great deal of pathos, only to proceed with a highly detached account of the "final solution," leaving the question of empathy with its victims to other, presumably Jewish historians.[21]

The German people did not want war. In 1939 most of them preferred things to remain as they had been during the preceding six years of Hitler's rule. Once war broke out, however, they found that much could be gained from it, and the Führer's popularity rose to even greater heights. Germans became anxious when the Wehrmacht invaded the Soviet Union. They did not want that war either. But while at first they expected Hitler to bring them world domination and wealth, later on they came to view him as their only hope for salvation. Instead, he brought them catastrophe. With this, they were released not only from his hold, but also from his crimes. They became his, and his victims' victims. Only now, while these pages are being written, are the Germans liberating themselves from the consequences of the catastrophe they had done so much to bring upon themselves. Ironically, their return to greatness is once more closely tied to Russia's retreat to the East. As the war generation is dying out, a new generation is being born into a reunified Germany. Let us hope that fifty years from now the new superpower emerging in *Mitteleuropa* will have an easier time remembering its past than the two rapidly vanishing German republics have had remembering theirs.

Notes

Introduction

1. See the most recent survey of this literature in T. Schulte, *The German Army and Nazi Policies in Occupied Russia* (Oxford/New York/Munich, 1989), pp. 1–27.

2. K.-J. Müller, *The army, politics and society in Germany, 1933–45* (New York, 1987); W. Deist, *The Wehrmacht and German Rearmament* (London/Basingstoke, 1981).

3. On Hitler, Nazism, and modernization, see, e.g., R. Zitelmann, *Hitler: Selbstverständnis eines Revolutionärs* (Stuttgart, 1987); H. A. Turner, Jr., "Fascism and Modernization," in *Reappraisals of Fascism*, ed. H. A. Turner, Jr. (New York, 1975), pp. 117–39; R. Dahrendorf, *Society and Democracy in Germany* (London, 1968); D. Schoenbaum, *Hitler's Social Revolution* (New York/London, 1966).

4. See, e.g., B. A. Carroll, *Design for Total War* (The Hague, 1968).

5. E. A. Shils and M. Janowitz, "Cohesion and Disintegration in the Wehrmacht in World War II," *POQ* 12 (1948): 280–315.

6. See, e.g., I. Kershaw, *Popular Opinion and Political Dissent in the Third Reich* (Oxford, 1983).

7. I. Kershaw, "How Effective Was Nazi Propaganda?," in *Nazi Propaganda*, ed. D. Welch (London, 1983), p. 192.

8. M. Messerschmidt, "German Military Law in the Second World War," in *The German Military in the Age of Total War*, ed. W. Deist (Leamington Spa/New Hampshire, 1985), pp. 323–35.

9. D. J. K. Peukert, "Alltag und Barbarei," in *Ist der Nationalsozialismus Geschichte?* ed. D. Diner (Frankfurt/M., 1987), pp. 51–61.

10. M. Hirsch et al. (eds.), *Recht, Verwaltung und Justiz im Nationalsozialismus* (Köln, 1984).

11. E. Klee, *"Euthanasie" im NS Staat. Die "Vernichtung lebensunwerten Lebens"* (Frankfurt/M., 1983); idem., *Dokumente zur "Euthanasie"* (Frankfurt/M., 1985); and idem., *Was sie taten—was sie wurden. Ärtze, Juristen und andere Beteiligte am Kranken—oder Judenmord* (Frankfurt/M., 1986); L. Dawidowicz, *The War against the Jews 1933–45*, 3d ed. (Harmondsworth, 1979), esp. pp. 78–101, 122–43; U. D. Adam, *Judenpolitik im Dritten Reich* (Düsseldorf, 1972); K. A. Schleunes, *The Twisted Road to Auschwitz. Nazi Policy toward German Jews, 1933–1939* (Urbana/Chicago/London, 1970); L. Gruchmann, " 'Blutschutzgesetz' und Justiz. Zur Entstehung und Auswirkung des Nürnberger Gesetzes vom 15. September 1935," *VfZ* 31 (1983): 418–42; O. D. Kulka, "Die Nürnberger Rassengesetze und die deutsche Bevölkerung im Lichte geheimer NS-Lage- und Stimmungsberichte," *VfZ* 32 (1984): 582–624.

12. A. Hillgruber, *Zweierlei Untergang* (Berlin, 1986); and textual analysis in O. Bartov, "Historians on the Eastern Front: Andreas Hillgruber and Germany's Tragedy," *TAJB* 16 (1987): 325–45.

13. J. Förster and G. R. Ueberschär, "Freiwillige für den 'Kreuzzug Europas gegen den Bolschewismus,' " in *Der Angriff auf die Sowjetunion* (Stuttgart, 1983), vol. 4 of *Das Deutsche Reich und der Zweite Weltkrieg*, 908–935; B. Wegner, "Auf dem Wege zur Pangermanischen Armee," *MGM* 2 (1980): 101–36.

14. Most of the relevant contributions are collected in *"Historikerstreit." Die Dokumentation der Kontroverse um die Einzigartigkeit der nationalsozialistischen Judenvernichtung*, ed. Serie Piper, 3d ed. (München, 1987).

15. M. Broszat, "Plädoyer für eine Historisierung des Nationalsozialismus," *Merkur* 435 (1985): 373–85.

16. For a more detailed exposition of this thesis, see also O. Bartov, "Extremfälle der Normalität und die Normalität des Aussergewöhnlichen: Deutsche Soldaten an der Ostfront," in *Über Leben im Krieg*, ed. U. Borsdorf and M. Jamin (Reinbeck bei Hamburg, 1989), pp. 148–61.

17. G. Hirschfeld (ed.), *The Policies of Genocide. Jews and Soviet Prisoners of War in Nazi Germany* (London/Boston/Sydney, 1986).

18. O. Bartov, *The Eastern Front 1941–45, German Troops and the Barbarisation of Warfare* (London/New York, 1985/6), pp. 40–67; Schulte, *German Army*, p. 288, and sources quoted therein.

19. See M. Geyer, "The Militarization of Europe, 1914–1945," in *The Militarization of the Western World*, ed. J. R. Gillis (New Brunswick/London, 1989), pp. 65–102.

Chapter 1

1. On the debate concerning Germany's economic preparations for and conduct of war, see A. S. Milward, *The German Economy at War* (London, 1965); T. W. Mason, "Innere Krise und Angriffskrieg," in *Wirtschaft und Rüstung am Vorabend des Zweiten Weltkrieges*, ed. F. Forstmeier and H.-E. Volkmann (Düsseldorf, 1975); H.-E. Volkmann, "Die NS-Wirtschaft in Vorbereitung des

Krieges," in *Ursachen und Voraussetzungen der deutschen Kriegspolitik* (Stuttgart, 1979), vol. 1 of *Das Deutsche Reich und der Zweite Weltkrieg*, 177–368; D. Kaiser, *Economic Diplomacy and the Origins of the Second World War* (Princeton: N.J., 1980); R. J. Overy, "Hitler's War and the German Economy: A Reinterpretation," *EHR* 35 (1982): 272–91; idem., "Germany, 'Domestic Crisis' and War in 1939," *P&P* 116 (1987): 138–68; "Debate: Germany, 'Domestic Crisis' and War in 1939," comments by D. Kaiser and T. W. Mason, reply by R. J. Overy, *P&P* 122 (1989): 200–40. Also see A. Speer, *Inside the Third Reich*, 5th ed. (London, 1979), esp. pp. 269–440.

2. On the tension between irrationality and fascination with technology which characterized Nazi ideology and some of its forerunners, see J. Herf, *Reactionary Modernism*, 2d ed. (Cambridge, 1986).

3. H. Umbreit, "Der Kampf um die Vormachtstellung in Westeuropa," in *Die Errichtung der Hegemonie auf dem Europäischen Kontinent* (Stuttgart, 1979), vol. 2 of *Das Deutsche Reich und der Zweite Weltkrieg*, 268, 282 (for figures), 284–307 (for an account of the campaign from the German side).

4. For the figures see ibid., pp. 254, 282. For a German Panzer commander's account of the campaign, see H. Guderian, *Panzer Leader*, 3d ed. (London, 1977), pp. 89–139. For the Allied angle see L. F. Ellis, *The War in France and Flanders, 1939–40* (London, 1953); A. Goutard, *The Battle of France, 1940* (London, 1958); J. Benoist-Méchin, *Sixty Days that Shook the West* (New York, 1963); G. Chapman, *Why France Collapsed* (London, 1968); A. Horne, *To Lose a Battle* (Harmondsworth, 1979). For two fascinating personal accounts, see M. Bloch, *Strange Defeat. A Statement of Evidence Written in 1940* (New York and London, 1968); H. Habe, *A Thousand Shall Fall* (London, 1942). Just as noteworthy are the letters collected in J.-P. Sartre, *Les carnets de la drôle de guerre* (Paris, 1983); and the fictional treatment in M. Tournier, *The Ogre* (New York, 1972).

5. R.-D. Müller, "Von der Wirtschaftsallianz zum kolonialen Ausbeutungskrieg," in *Das Deutsche Reich und der Zweite Weltkrieg*, 4:183. See also H. Schustereit, *Vabanque: Hitlers Angriff auf die Sowjetunion 1941 als Versuch, durch den Sieg im Osten den Westen zu bezwingen* (Herford, F.R.G., 1988).

6. Müller, *Wirtschaftsallianz*, pp. 183–85; J. Hoffmann, "Die Sowjetunion bis zum Vorabend des deutschen Angriffs," in *Das Deutsche Reich und der Zweite Weltkrieg*, 4:62–75; J. Hoffmann, "Die Kriegführung aus der Sicht der Sowjetunion," in *Das Deutsche Reich und der Zweite Weltkrieg*, 4:734; J. Erickson, *The Road to Stalingrad*, 2d ed. (London, 1985), vol. 1 of *Stalin's War with Germany*, pp. 93, 322; Ploetz, *Geschichte des Zweiten Weltkrieges*, 2d ed. (Würzburg, 1960), pp. 122–27.

7. Ploetz, *Geschichte*, pp. 448–53, 471, 499, 593–94, 613; Hoffmann, *Die Sowjetunion*, p. 734.

8. See the following contributions in vol. 4 of *Das Deutsche Reich und der Zweite Weltkrieg*: J. Förster, "Hitlers Entscheidung für den Krieg gegen die Sowjetunion," pp. 3–37; E. Klink, "Die militärische Konzeption des Krieges gegen die Sowjetunion: Die Landkriegführung," pp. 190–277; idem., "Die Operationsführung: Heer und Kriegsmarine," pp. 451–652; R.-D. Müller, "Das Scheitern der

wirtschaftlichen 'Blitzkriegstrategie,' " pp. 936–1029. Also see M. van Creveld, *Supplying War. Logistics from Wallenstein to Patton*, 3d ed. (New York, 1980), pp. 142–80; H. Rohde, *Das Deutsche Wehrmachttransportwesen im Zweiten Weltkrieg* (Stuttgart, 1971); W. Zieger, *Das deutsche Heeresveterinärwesen im Zweiten Weltkrieg* (Freiburg i.Br., 1973).

    9. Müller, *Das Scheitern*, pp. 965–67.

    10. Bundesarchiv-Militärarchiv, Freiburg i.Br. (henceforth BA-MA), RH26–12/30, 22.6.–21.7.41; BA-MA, RH26–12/27, 22.6.–6.7.41; BA-MA, RH37/943, *Kriegstagebuch*, 13.6.–13.8.41.

    11. BA-MA, RH26–12/38, 28.10.41.

    12. Klink, *Die Operationsführung*, pp. 632–34; A. Seaton, *The Russo-German War 1941–45* (London, 1971), pp. 230–48; D. Irving, *The Rise and Fall of the Luftwaffe* (London, 1976), pp.183–97.

    13. BA-MA, RH26–12/212, 21.1.42; BA-MA, RH26–12/61, 18.2.42; BA-MA, RH26–12/62, 4.3.42 and 7.3.42; BA-MA, RH26–12/142, 24.1.42, 9.5.42; BA-MA, RH26–12/52, 1.4.–15.5.42.

    14. BA-MA, RH26–12/142, 24.1.42.

    15. BA-MA, RH26–12/63, 22.4.42.

    16. BA-MA, RH26–12/53, 25.6.42.

    17. Müller, *Das Scheitern*, pp. 967–89.

    18. W. Paul, *Geschichte der 18. Panzer Division 1940–43* (Freiburg i.Br., n.d.), pp. 10, 30–31, 35. Also see Guderian, *Panzer Leader*, pp. 159–67.

    19. BA-MA, RH27–18/17, 11.7.41.

    20. Paul, *18.Pz.Div.*, p. 46.

    21. Ibid., pp. 36–38.

    22. BA-MA, RH27–18/26, 27.7.41. Generally on the uses and abuses of psychology in the Third Reich, including the Wehrmacht, see U. Geuter, *Die Professionalisierung der deutschen Psychologie im Nationalsozialismus*, 2d ed. (Frankfurt/M., 1988).

    23. Paul, *18.Pz.Div.*, pp. 51–52.

    24. Ibid., pp. 64–67; BA-MA, RH27–18/27, 14.9.41.

    25. O. Buchbender and R. Stertz (eds.), *Das andere Gesicht des Krieges* (München, 1982), pp. 81–82, letter 127.

    26. Paul, *18.Pz.Div.*, pp. 110, 125.

    27. BA-MA, RH27–18/63, 10.12.41 for 5.12.41.

    28. BA-MA, RH27–18/73, 9.11.41; BA-MA, RH27–18/177, 23.–30.11.41; 7.12.41; BA-MA, RH27–18/74, 2.12.41; Paul, *18.Pz.Div.*, pp. 131–34. For a vivid personal account of one soldier's misery during the march toward Moscow, see letter cited in W. Bähr and H. W. Bähr (eds.), *Kriegsbriefe gefallener Studenten 1939–1945* (Tübingen/Stuttgart, 1952), pp. 13–14.

    29. BA-MA, RH27–18/69, 6.1.42 for 1.1.42; BA-MA, RH27–18/75, 6.2.42; BA-MA, RH27–18/63, 22.2.42 for 15.2.42.

    30. Paul, *18.Pz.Div.*, p. 149.

    31. Ibid., p. 152.

    32. Buchbender/Stertz, *Das andere Gesicht*, pp. 92–93, letter 156.

33. BA-MA, RH27–18/184, 7.1.–31.3.42.

34. See documents cited *infra*, note 28. Also BA-MA, RH27–18/69, 28.2.42; BA-MA, RH27–18/113, 19.3.42 for 18.3.42; BA-MA, RH27–18/69, 28.3.42. Even in April 1943 out of 31 available tanks 8 were obsolete Panzer IIs, 3 were command tanks without a gun, and 8 were in repairs; of 29 anti-tank guns, 10 were the useless 3.7cm model, and 10 were Russian booty, lacking ammunition and spare parts. See BA-MA, RH27–18/130, 7.4.43.

35. BA-MA, RH27–18/184, 8.3.42.

36. See, e.g., BA-MA, RH26–12/58, 1.1.43.

37. BA-MA, RH27–18/74, 4.1.42.

38. BA-MA, RH27–18/74, 15.1.42.

39. BA-MA, RH27–18/63, 22.2.42 for 15.2.42.

40. BA-MA, RH27–18/174, 1.1.–31.3.42.

41. B. R. Kroener, "Die Personellen Ressourcen des Dritten Reiches im Spannungsfeld zwischen Wehrmacht, Bürokratie und Kriegswirtschaft 1939–1942," in *Organisation und Mobilisierung des deutschen Machtbereichs*, vol. 5/1 of *Das deutsche Reich und der Zweite Weltkrieg* (Stuttgart, 1988), 872–85. See also idem., "Squaring the Circle. Blitzkrieg Strategy and the Manpower Shortage, 1939–42," in *The German Military in the Age of Total War*, ed. W. Deist (Leamington Spa, New Hampshire, 1985), pp. 282–303; and F. Seidler, *Prostitution, Homosexualität, Selbstverstümmelung. Probleme der deutsche Sanitätsführung 1939–45* (Neckargemünd, 1977), p. 33. But note G. Cocks, "The Professionalization of Psychotherapy in Germany, 1928–1949," in *German Professions, 1800–1950*, ed. G. Cocks and K. Jarausch (New York, 1990), esp. p. 327, n. 68.

42. BA-MA, RH27–18/115, 9.4.42.

43. K. Reinhardt, *Die Wende vor Moskau* (Stuttgart, 1972), pp. 212–15; Kroener, *Die Personellen Ressourcen*, pp. 923–24.

44. On the machine-gun see, e.g., J. Ellis, *The Social History of the Machine Gun*, 2d ed. (London, 1987); on the image of war in 1914–18 see E. J. Leed, *No Man's Land* (Cambridge, 1979); more generally on the image of war see O. Bartov, "Man and the Mass: Reality and the Heroic Image in War," *H&M* 1 (1989): 99–122.

45. Bähr, *Kriegsbriefe*, p. 74.

46. Buchbender/Stertz, *Das andere Gesicht*, p. 78, letter 114.

47. Bähr, *Kriegsbriefe*, p. 76.

48. H. Spaeter and W. Ritter von Schramm, *Die Geschichte des Panzerkorps Grossdeutschland* (Bielefeld, 1958), 1:341–46.

49. Bähr, *Kriegsbriefe*, pp. 159–60.

50. Ibid., pp. 302–3.

51. Spaeter/Schramm, *GD*, 1:365–66. Self-preservation is also the leitmotif of a letter sent by Karl Fuchs from the Eastern Front on 4 August 1941. Cited in H. F. Richardson (ed), *Sieg Heil! War Letters of Tank Gunner Karl Fuchs 1937–1941* (Hamden: Conn., 1987), p. 124.

52. Ibid., pp. 13–14.

53. Ibid., p. 157. On the role played by the myth of the war experience of

the First World War in interwar Germany and its mobilization by the Nazis, see
G. L. Mosse, *Fallen Soldiers. Reshaping the Memory of the World Wars* (New
York, 1990), esp. pp. 159–200.

54. Spaeter/Schramm, *GD*, 1:366–68.

55. Bähr, *Kriegsbriefe*, pp. 83–84.

Chapter 2

1. On the U.S. Army versus the Wehrmacht, see M. van Creveld, *Fighting
Power* (Westport: Conn., 1982); further on the German tradition see, e.g., D.
Bald, *Der deutsche Offizier* (München, 1984); F. L. Carsten, "Germany. From
Scharnhorst to Schleicher: the Prussian Officer Corps in Politics, 1806–1933," in
Soldiers and Governments, ed. M. Howard (London, 1957); G. A. Craig, *The
Politics of the Prussian Army, 1640–1945,* 3d ed. (London, 1978); K. Demeter, *The
German Officer Corps in Society and State, 1650–1945* (London, 1965); M. Kitchen,
The German Officer Corps, 1890–1914 (London, 1968); H. Kurze, "Das Bild des
Offiziers in der deutschen Literatur," in *Das deutsche Offizierkorps, 1860–1960,*
ed. H. H. Hofmann (Boppard am Rhein, 1980), pp. 413–35; M. Messerschmidt,
"Werden und Prägung des preussischen Offizierkorps—Ein Überblick," in *Offi-
ziere im Bild vom Dokumente aus drei Jahrhunderten* (Stuttgart, 1964), pp. 11–
104. A good British equivalent is the "Pals" Battalions. See, e.g., J. Keegan, *The
Face of Battle,* 2d ed. (London, 1976), pp. 215–25; J. M. Winter, *The Great War
and the British People* (Cambridge: Mass., 1986), pp. 25–39; J. Stevenson, *British
Society 1914–45* (Harmondsworth, 1984), pp. 46–57. On the French debate re-
garding the appropriate conscription system see, e.g., R. D. Challener, *The French
Theory of the Nation in Arms, 1866–1939* (New York, 1955); P. C. F. Bankwitz,
"Maxime Weygand and the Army-Nation Concept in the Modern French Army,"
FHS 2 (1961): 157–88; and on the origins of the modern French concept: I. Woloch,
"Napoleonic Conscription: State Power and Civil Society," paper delivered at the
Davis Center Seminar, Princeton University, January 6, 1984; J.-P. Bertaud, "Na-
poleon's Officers," *P&P* 112 (1986): 91–111; D. D. Bien, "The Army in the French
Enlightenment: Reform, Reaction and Revolution," *P&P* 85 (1979): 68–98.

2. Shils/Janowitz, *Cohesion and Disintegration,* p. 281; in the same vein, see
also M. I. Gurfein and M. Janowitz, "Trends in Wehrmacht Morale," in *Propaganda in War and Crisis,* ed. D. Lerner (New York, 1951), pp. 200–208.

3. Ibid.

4. See, e.g., E. P. Chodoff, "Ideology and Primary Groups," *AFS* 9 (1983):
569–93. More generally on combat motivation see R. Holmes, *Firing Line,* 2d ed.
(Harmondsworth, 1987); A. Kellet, *Combat Motivation* (Boston, 1982); S. L. A.
Marshall, *Men against Fire* (New York, 1947); Lord Moran, *The Anatomy of
Courage,* 2d ed. (London, 1966); F. M. Richardson, *Fighting Spirit* (London, 1978).

5. van Creveld, *Fighting Power,* p. 166.

6. Ibid., pp. 163–64. See also, e.g., debates regarding this issue in the journal
of the Israeli Defense Forces: Gen. A. Tamir, "Quality Versus Quantity,"

Ma'arakhot 250 (1976): 8–12, 38; Lt.-Col. (res.) Y. Gelber, "Military Education and Ideology," *Ma'arakhot* 267 (1979): 8–12 (both in Hebrew).

7. For a discussion of oral history, see P. Thompson, *The Voice of the Past: Oral History* (Oxford, 1978); more specifically on the oral history of the Third Reich, see L. Niethammer, (ed.), *"Die Jahre weiss man nicht, wo man die heute hinsetzen soll." Faschismuserfahrungen im Ruhrgebiet* (Berlin, Bonn, 1983); on the related issue of so-called *Alltagsgeschichte* see, e.g., M. Broszat and E. Fröhlich, *Alltag und Widerstand* (München, 1987); Kershaw, *Popular Opinion*; L. Niethammer, "Anmerkungen zur Alltagsgeschichte," in *Geschichte im Alltag—Alltag in der Geschichte*, ed. K. Bergmann and R. Schörken (Düsseldorf, 1982), pp. 11–29; K. Tenfelde, "Schwierigkeiten mit dem Alltag," *GuG* 10 (1984): 376–94; W. Wippermann, "Fascism and the History of Everyday Life," paper delivered at the *Wiener Library Seminar*, Tel-Aviv University, 1987.

8. For a brief discussion of this issue see O. Bartov, "Daily Life and Motivation in War: The Wehrmacht in the Soviet Union," *JSS* 12 (1989): 200–14.

9. Richardson, *Sieg Heil!*, p. 147. See also M. Messerschmidt, "The Wehrmacht and the Volksgemeinschaft," *JCH* 18 (1983): 719–40.

10. Bähr, *Kriegsbriefe*, p. 150.

11. Ibid., pp. 90–91.

12. For a recent legitimation of this view see Hillgruber, *Zweierlei Untergang*.

13. Shils/Janowitz, *Cohesion and Disintegration*, p. 301.

14. Kroener, *Die Personellen Ressourcen*, pp. 871–78. Generally on the expansion of the Wehrmacht during the 1930s and on the eve of "Barbarossa," see G. Tessin, *Formationsgeschichte der Wehrmacht, 1933–39* (Boppard am Rhein, 1959), pp. 21–22, 28–29, 114–15, 118–19, 150–51, 174–75, 190–91; B. Mueller-Hillebrand, *Das Heer*, 3 vols (Darmstadt, 1954; Frankfurt/M., 1956 and 1969), 1:15, 55–59, 68–73, 130–35, 151; W. Deist, "Die Aufrüstung der Wehrmacht," in *Ursachen und Voraussetzungen der deutschen Kriegspolitik* (Stuttgart, 1979), vol. 1 of *Das Deutsche Reich und der Zweite Weltkrieg*, 415–49, esp. 447–48; H. Rohde, "Hitlers erster 'Blitzkrieg' und seine Auswirkungen auf Nordosteuropa," in *Das Deutsche Reich und der Zweite Weltkrieg*, 2:99–104; H. G. Dahms, *Die Geschichte des Zweiten Weltkrieges* (München, Berlin, 1983), p. 88; B. R. Kroener, "Auf dem Weg zu einer 'nationalsozialistischen Volksarmee.' Die soziale Öffnung des Heeresoffizierkorps im Zweiten Weltkrieg," in *Von Stalingrad zur Währungsreform. Zur Sozialgeschichte des Umbruchs in Deutschland*, ed. M. Broszat et al. (München, 1988), pp. 651–82.

15. Kroener, *Die Personellen Ressourcen*, pp. 884–94, 913–16, 920–22.

16. On the larger economic and political limitations on total mobilization see *infra*, Chapter 1, note 1. Also see, on women's economic mobilization: T. W. Mason, "Women in Germany, 1925–40: Family, Welfare and Work," Parts 1–2, *HWJ* 1 (1976): 74–113; 2 (1976): 5–32; S. Dammer, "Kinder, Küche, Kriegsarbeit: Die Schulung der Frauen durch die NS-Frauenschaft," in *Mutterkreuz und Arbeitsbuch*, ed. Frauengruppe Faschismusforschung (Frankfurt/M., 1981), pp. 215–45; R. Bridenthal, *Mothers in the Fatherland* (New York, 1987); idem. et al. *When Biology Became Destiny* (New York, 1984); S. J. McIntyre, "Women and the

Professions in Germany 1930–40," in *German Democracy and the Triumph of Hitler,* ed. A. Nicholls and E. Matthias (London, 1971), pp. 175–213; J. Stephenson, *Women in Nazi Society* (London, 1975), and idem., " 'Emancipation' and its Problems: War and Society in Württemberg 1939–45," *EHQ* 17 (1987): 345–65; D. Winkler, *Frauenarbeit im "Dritten Reich"* (Hamburg, 1977). On workers, see T. W. Mason (ed.), *Arbeiterklasse und Volksgemeinschaft* (Opladen, 1975); idem., *Sozialpolitik im Dritten Reich* (Opladen, 1977); D. Petzina, "Die Mobilisierung Deutscher Arbeitskräfte vor und während des Zweiten Weltkrieges," *VfZ* 4 (1970): 443–55. On planning, R. J. Overy, *Goering: The "Iron Man"* (London, 1984). On foreign workers, U. Herbert, *Fremdarbeiter: Politik und Praxis des "Ausländer-Einsatzes" in der Kriegswirtschaft des Dritten Reiches* (Berlin and Bonn, 1985).

17. Kroener, *Die Personellen Ressourcen,* pp. 922–23, 926–28.

18. F. Halder, *Kriegstagebuch 1939–42,* ed. H.-A. Jacobsen, 3 vols. (Stuttgart, 1962–64), 3:345, 418; *Kriegstagebuch des Oberkommando der Wehrmacht 1940–45,* ed. P. E. Schramm, 4 vols. (Frankfurt/M., 1961–65), 1:1120–21.

19. Kroener, *Die Personellen Ressourcen,* pp. 878–81, 924–25.

20. Ibid., pp. 894–906. On changing criteria for promotion, see G. Papke, "Offizierkorps und Anciennität," in *Untersuchungen zur Geschichte des Offizierkorps,* ed. H. Meier-Welcker (Stuttgart, 1962), pp. 177–206. Further figures on officer casualties in Halder, *Kriegstagebuch,* 3:345; *KTB-OKW,* 1:1120–21. More detailed accounts in J. Flottmann and H. Möller-Witten, *Opfergang der Generale. Die Verluste der Generale und Admirale und der im gleichen Dienstrang stehenden sonstigen Offiziere und Beamten im Zweiten Weltkrieg* (Berlin, 1952); R. Stumpf, *Die Wehrmacht-Elite. Rang- und Herkunftsstruktur der deutschen Generale und Admirale 1933–45* (Boppard am Rhein, 1982).

21. Paul, *18.Pz.Div.,* pp. 1–4, 10, 14–15; BA-MA, RH27–18/2a, 26.9.40; BA-MA, RH27–18/1, 27.5.41; BA-MA, RH27–18/3, 15.3.41.

22. BA-MA, RH27–18/26, 22.6.–10.7.41.

23. Paul, *18.Pz.Div.,* pp. 39–43; BA-MA, RH27–18/26, 27.7.41.

24. Paul, *18.Pz.Div.,* pp. 43–48, 50–53.

25. Spaeter/Schramm, *GD,* 1:266–70.

26. Ibid., pp. 276–79, 280–302, 304–308.

27. BA-MA, RH27–18/69, 1.10.41.

28. BA-MA, RH27–18/63, 5.11.41 for 31.10.41; BA-MA, RH27–18/69, 10.11.41 for 1.11.41.

29. Paul, *18.Pz.Div.,* p. 134.

30. Spaeter/Schramm, *GD,* 1:347, 355–59.

31. BA-MA, RH26–12/46, 6.11.41.

32. BA-MA, RH26–12/22, 1.6.–10.12.41; BA-MA, RH26–12/130, 10.12.41.

33. BA-MA, RH27–18/74, 22.12.41, 27.12.41.

34. Reinhardt, *Die Wende,* pp. 206–11.

35. BA-MA, RH27–18/174, 1.1.-31.3.42.

36. Paul, *18.Pz.Div.,* pp. 19–21, 202; BA-MA, RH27–18/174, 1.1.-31.3.42, 21.6.-31.12.41; BA-MA, RH27–18/60, 15.12.41–22.1.42.

37. BA-MA, RH26–12/63, 22.4.42; BA-MA, RH26–12/64, 23.4.42.

38. BA-MA, RH26–12/92, 16.12.41–28.2.43; BA-MA, RH26–12/52, 4.4.42, 25.4.42.

39. Spaeter/Schramm, *GD*, 1:385.

40. Ibid., pp. 386–400.

41. Bähr, *Kriegsbriefe*, p. 90.

42. Ibid., pp. 212–13.

43. Ibid., p. 100.

44. Mueller-Hillebrand, *Das Heer*, 3:51–3; Dahms, *Geschichte*, p. 362; Halder, *Kriegstagebuch*, 3:426–27, 430–32, 465.

45. Halder, *Kriegstagebuch*, p. 522.

46. *KTB OKW*, 3 (part 2):1481–44; Mueller-Hillebrand, *Das Heer*, 2:109ff; 3:111–14, 132–35; E. Klink, *Das Gesetz des Handelns* (Stuttgart, 1966), p. 34; Dahms, *Geschichte*, pp. 439–40; J. Hoffmann, *Die Ostlegionen 1941–43* (Freiburg i.Br., 1977); idem., *Deutsche und Kalmyken 1942 bis 1945* (Freiburg i.Br., 1977); P. H. Buss, "The Non-Germans in the German Armed Forces 1939–45" (Canterbury Univ. Ph.D. thesis, 1974). On similar developments in the Waffen-SS see Wegner, *Pangermanische Armee*; and idem., "Das Führerkorps der Waffen-SS im Kriege," in *Das deutsche Offizierkorps, 1860–1960*, ed. H. H. Hofmann (Boppard am Rhein, 1980), pp. 327–50.

47. Dahms, *Geschichte*, p. 478; Mueller-Hillebrand, *Das Heer*, 3:136–37.

48. *KTB OKW*, 4 (part 2):1509–11.

49. H. Magenheimer, *Abwehrschlachten an der Weichsel* 1945 (Freiburg i.Br., 1976), pp. 45–47, according to whom the monthly average of casualties during the second half of 1944 was 198,986 men, of whom 20,567 were killed; Dahms, *Geschichte*, p. 548.

50. *KTB OKW*, 4 (part 2):1508–11, 1514–16.

51. BA-MA, RH27–18/117, 25.4.42 for 19.4.42; 20.4.42; 25.4.42; 20.6.42 for 15.6.42.

52. Paul, *18.Pz.Div.*, pp. 209–24.

53. Ibid., p. 222.

54. BA-MA, RH27–18/131, 12.3.42.

55. BA-MA, RH27–18/130, 7.4.43.

56. BA-MA, RH26–12/265, 22.5.42.

57. BA-MA, RH26–12/53, 10.5.42.

58. BA-MA, RH26–12/53, 5.6.42; BA-MA, RH26–12/92, 16.12.41–28.2.43.

59. BA-MA, RH26–12/266, 20.7.42.

60. BA-MA, RH26–12/66, 12.7.42, 17.7.42, 30.7.42.

61. BA-MA, RH26–12/55, 18.8.42.

62. BA-MA, RH26–12/55, 19.8.42.

63. BA-MA, RH26–12/56, 26.10.42.

64. BA-MA, RH26–12/57, 16.11.42; BA-MA, RH26–12/69, 21.11.42, 26.11.42, 6.12.42; BA-MA, RH26–12/49, *Kriegstagebuch*, 16.12.41–18.2.43, entries for 25.11.42 and 29.12.42; BA-MA, RH26–12/58, 14.12.42.

65. BA-MA, RH26–12/57, 25.11.42, 8.12.42; BA-MA, RH26–12/69, 26.11.42, 6.12.42.

66. BA-MA, RH26–12/57, 8.12.42, 10.12.42.

67. BA-MA, RH26–12/58, 1.1.43; BA-MA, RH26–12/92, 16.12.41–28.2.43.

68. BA-MA, RH26–12/75, 1.3.-21.6.43.

69. BA-MA, RH26–12/76, 23.3.43.

70. BA-MA, RH26–12/49, *Kriegstagebuch*, 16.12.41–28.2.43; *Verlustliste*: 16.12.41–30.4.43; BA-MA, RH26–12/265, 16.10.43.

71. BA-MA, RH26–12/167, accounts of the Vitebsk battles.

72. BA-MA, *Heft* 523, pp. 12–13.

73. Spaeter/Schramm, *GD*, 1:414, 429, 548–52; BA-MA, RH26–1005/5, 27.6.-31.7.42, 27.6.-18.8.42, and later reports in same box; BA-MA, RH26–1005/7, 1.7.-31.7.42, 1.8.42; BA-MA, RH26–1005/60, 15.5.-31.7.42.

74. Spaeter/Schramm, *GD*, 1:552–606; RH37/6337, *Tagesbericht C. Benes*, p. 13.

75. Benes, pp. 13–14.

76. BA-MA, RH26–1005/78, *Tagebuch von Hobe*, pp. 14–21.

77. BA-MA, RH37/6351, *Bericht F. Mebes*, entry for 9.12.42, and more generally pp. 22–26. Detailed account of the fighting in Spaeter/Schramm, *GD*, 1: 606–55.

78. Mebes, entry for 31.12.42; Spaeter/Schramm, *GD*, 1:655–67.

79. Spaeter/Schramm, *GD*, 1:667–69.

80. BA-MA, RH37/6394, 9.1.43.

81. BA-MA, RH26–1005/50, 26.4.43, 1.8.42–25.4.43 (troop casualties), and 1.8.42–25.4.43 (officer casualties).

82. Bähr, *Kriegsbriefe*, p. 170.

83. Ibid., p. 172.

84. BA-MA, RH27–18/196, 30.4.43.

85. BA-MA, RH27–18/169, 1.4.-30.6.43; BA-MA, RH27–18/131, 7.5.43.

86. BA-MA, RH27–18/164, 30.6.43.

87. BA-MA, RH27–18/144, 10.7.43 and 11.7.43; Paul, *18.Pz.Div.*, p. 248.

88. BA-MA, RH27–18/144, 20.7.43 for 11.7.43, 21.7.43, 26.7.43 for 21.7.43; Paul, *18.Pz.Div.*, p. 268.

89. BA-MA, RH27–18/144, 23.7.43; BA-MA, RH27–18/142, 24.7.43.

90. BA-MA, RH27–18/144, 19.7.43, 27.7.43; Paul, *18.Pz.Div.*, pp. 268–70.

91. BA-MA, RH27–18/144, 1.8.43, 11.8.43.

92. BA-MA, RH27–18/170, 3.8.43; BA-MA, RH27–18/144, 20.8.43 for 11.8.43, 25.8.43, 1.9.43; Paul, *18.Pz.Div.*, pp. 273–74.

93. BA-MA, RH27–18/144, 17.9.43, 1.10.43.

94. BA-MA, RH27–18/170, 1.10.43. These figures do not include the sick, who in winter 1941–42 alone numbered close to 5000, nor the thousands of men suffering from various degrees of frostbite.

95. For a discussion of other factors contributing to the soldiers' fighting ferocity and the absence of mutinies, such as physical isolation, cultural determinants and differences, the bombing of German cities, and so forth, see O. Bartov,

"The Conduct of War: Soldiers and the Barbarization of Warfare," *JMH, Special Issue: Resistance Against the National Socialist Regime* (forthcoming 1991).

96. Spaeter/Schramm, *GD*, 2:167–214; Benes, p. 17.

97. Spaeter/Schramm, *GD*, 2:251–70; Benes, pp. 18–21.

98. Spaeter/Schramm, *GD*, 2:270–94; Benes, pp. 21–22.

99. BA-MA, RH26–1005/51, 8.11.43, 1.4.–30.9.43 (troop casualties), and 1.4.–30.9.43 (officer casualties).

100. Spaeter/Schramm, *GD*, 2:329–99.

101. Ibid., 409–17.

102. Benes, pp. 23–28. Also see BA-MA, RH26–1005/82, *"Tapferkeitsdaten,"* 15.–17.10.43; BA-MA, RH26–1005/82, *"Kriegsberichterzug der Pz.Gren.Div. GD.,"* 2.11.43.

103. Bähr, *Kriegsbriefe*, p. 211.

104. Spaeter/Schramm, *GD*, 2:450–87; Benes, pp. 31–33.

105. Spaeter/Schramm, *GD*, 2:487–527.

106. BA-MA, RH26–1005/81, *"Sturmgeschütz Blatt;"* Spaeter/Schramm, *GD*, 2:527–44.

107. Ibid., 590–626.

108. BA-MA, RH37/6394, *"Stellenbesetzung I/Pz.Gren.Rgt.GD 6.10. bis 15.10.44."* Also see ibid., *"Notizen über Pz.Gren.Rgt.GD 13.6.-13.8.44, Stellenbesetzung III/Pz.Gren.Rgt.GD;"* BA-MA, RH26–1005/84, issue of the *Frontzeitung "Die Feuerwehr;"* BA-MA, RH26–1005/79, accounts of acts of heroism among divisional officers as grounds for decoration with the *Ritterkreuz*; Spaeter/Schramm, *GD*, 2:626–714.

109. Spaeter/Schramm, *GD*, 3:11–24.

110. Ibid., pp. 182–261, 381–404; BA-MA, RH26–1005/81, *"Major Helmut Kraussold,"* for a brief biographical account of one of the division's longest serving officers.

111. BA-MA, RH26–1005/82, *"Neu-Aufstellung der 'Pz.Gren.Div. Grossdeutschland'."* Spaeter/Schramm, *GD*, 3:656–65.

112. Benes, pp. 42–53; BA-MA, RH26–1005/82, *"Bericht zu dem Ereignissen vom 1.3.-18.4.45,"* written after the war; Spaeter/Schramm, *GD*, 3:665–81.

Chapter 3

1. Corporeal punishment for minor breaches of discipline was abolished only in 1808. See Craig, *Politics*, p. 48.

2. See, e.g., J. F. C. Fuller, *The Decisive Battles of the Western World*, 2d ed. (London, 1970), 1: 555–78; M. Howard, *The Franco-Prussian War*, 3d ed. (London, 1981), pp. 57–63, 77–85; B. H. Liddell Hart, *History of the First World War*, 3d ed. (London, 1979).

3. See, e.g., Craig, *Politics*, pp. 349–64; F. L. Carsten, *The Reichswehr and*

Politics 1918–33, 2d ed. (Berkeley/Los Angeles, 1973), pp. 3–48; K.-J. Müller, *General Ludwig Beck* (Boppard am Rhein, 1980), pp. 323–39.

4. See mainly Kroener, *Volksarmee*; Messerschmidt, *Werden und Prägung*; R. Absolon, "Das Offizierkorps des deutschen Heeres 1935–1945," in *Das deutsche Offizierkorps 1860–1960*, ed. H. H. Hofmann (Boppard am Rhein, 1980), pp. 247–68; Papke, *Offizierkorps und Anciennität*; J. Wheeler-Bennett, *The Nemesis of Power*, 2d ed. (London, 1980), pp. 332, 393–94. On similarities in linguistic usage, see V. Klemperer, *Die Unbewältigte Sprache*, 3d ed. (Darmstadt, n.d.); C. Berning, *Vom "Abstammungsnachweis" zum "Zuchtwart." Vokabular des Nationalsozialismus* (Berlin, 1964).

5. R. J. O'Neill, *The German Army and the Nazi Party, 1933–39* (London, 1966); M. Messerschmidt, *Die Wehrmacht im NS-Staat* (Hamburg, 1969); K.-J. Müller, *Das Heer und Hitler* (Stuttgart, 1969).

6. Messerschmidt, *German Military Law*. On "justice" toward the occupied in the East, see D. Majer, *"Fremdvölkische" im Dritten Reich* (Boppard am Rhein, 1981). Generally on the perversion of law in the Reich, see S. König, *Vom Dienst am Recht: Rechtsanwälte als Strafverteidiger im Nationalsozialismus* (New York, 1987); B. Rüthers, *Entartetes Recht: Rechtslehren und Kronjuristen im Dritten Reich* (München, 1988); D. Diner, "Rassistisches Völkerrecht: Elemente einer nationalsozialistischen Weltordnung," *VfZ* 1 (1989): 23–56; D. Güstow, *Tödlicher Alltag. Strafverteidiger im Dritten Reich* (Berlin, 1981); H. Focke and M. Strocka, *Alltag der Gleichgeschalteten* (Reinbek bei Hamburg, 1985), pp. 204–62.

7. Generally on the partisan campaign, see J. A. Armstrong (ed.), *Soviet Partisans in World War II* (Wisconsin, 1964); M. Cooper, *The Phantom War* (London, 1979); H. Kühnrich, *Der Partisanenkrieg in Europa, 1939–45* (Berlin, 1965); V. Redelis, *Partisanenkrieg* (Heidelberg, 1958).

8. On "Barbarossa" as a crusade see Förster/Ueberschär, *Freiwillige*; J. Förster, " 'Croisade de l'Europe contre le Bolchevisme,' " *RHDGM* 30 (1980): 1–26; A. J. Mayer, *Why Did the Heaven Not Darken?* (New York, 1988).

9. BA-MA, RH26–12/252, 25.10.39.

10. BA-MA, RH26–12/252, 20.11.39.

11. BA-MA, RH26–12/252, 18.12.39.

12. BA-MA, RH26–12/279, 29.9.39.

13. BA-MA, RH26–12/279, 1.10.39.

14. BA-MA, RH26–12/99, 25.10.40.

15. BA-MA, RH26–12/236, 8.11.39.

16. BA-MA, RH26–12/253, 19.9.39.

17. BA-MA, RH26–12/252, 25.10.39. On the army death penalty for homosexuality decreed in 1943, see Cocks, *Professionalization of Psychotherapy*, pp. 319–21. I am informed that Geoffrey Giles is currently working on a book on homosexuality in Nazi Germany. Also see Seidler, *Prostitution*, pp. 193–232.

18. Rohde, *Hitlers erster "Blitzkrieg,"* pp. 136–46. See also, M. Broszat, *Nationalsozialistische Polenpolitik 1939–45* (Stuttgart, 1961); H. Umbreit, *Deutsche Militärverwaltung 1938/39* (Stuttgart, 1977).

19. E. Klee et al. (eds.), *"Schöne Zeiten." Judenmord aus der Sicht der Täter und Gaffer* (Frankfurt/M., 1988), pp. 14–15.

20. C. Streit, *Keine Kameraden* (Stuttgart, 1978), pp. 187–88 notes that the Wehrmacht took great care to ensure the proper treatment of POWs in the Western campaign, measures which were quite unthinkable in Russia.

21. H. Umbreit, *Der Militärbefehlshaber in Frankreich 1940–44* (Boppard am Rhein, 1968); and idem., *Vormachtstellung*, pp. 319–27. See also, E. Jäckel, *Frankreich in Hitlers Europa* (Stuttgart, 1966); A. Hillgruber, *Hitlers Strategie. Politik und Kriegführung 1940–41* (Frankfurt/M., 1965). However, see C. W. Sydnor, Jr., *Soldiers of Destruction. The SS Death's Head Division, 1933–1945* (Princeton: N.J., 1977), pp. 106–9, on the massacre of British POWs in Le Paradis and the lack of measures by the Wehrmacht command against the culprits. On the impact of the experience in the East on formations transferred to the West in 1944, see M. Hastings, *Das Reich. Resistance and the March of the 2nd SS Panzer Division Through France June 1944* (London, 1981), for the massacre of Oradour, and especially his comment on p. 16; and J. J. Weingartner, *Crossroads of Death. The Story of the Malmédy Massacre and Trial* (Berkeley/Los Angeles, 1979). That there was an element of mythical and occasionally also real chivalry especially in the *Luftwaffe* and Navy, due to both different traditions and technical circumstances, in no way contradicts the fact that both services accepted the "necessity" to kill civilians, nor that they were deeply permeated by Nazi indoctrination. See, e.g., Kroener, *Die Personellen Ressourcen*, p. 912.

22. BA-MA, RH26–12/183, 21.5.40.

23. BA-MA, RH26–12/274, 27.6.40.

24. BA-MA, RH26–12/235, 3.10.40.

25. BA-MA, RH26–12/99, 25.10.40.

26. BA-MA, RH26–12/235, 2.10.40; BA-MA, RH26–12/108, 9.4.41.

27. BA-MA, RH26–12/21, 7.5.41.

28. BA-MA, RH26–12/21, 6.5.41.

29. BA-MA, RH26–12/21, 8.5.41.

30. BA-MA, RH26–12/236, 21.6.40.

31. Streit, *Keine Kameraden*, pp. 28–61. See also H. Krausnick and H.-H. Wilhelm, *Die Truppe des Weltanschauungskrieges* (Stuttgart, 1981), pp. 116–41; H.-A. Jacobsen, "Kommissarbefehl und Massenexekutionen sowjetische Kriegsgefangener," in *Anatomie des SS-Staates*, H. Buchheim et al. (Olten, 1965), 2:170–82; H. Krausnick, "Kommissarbefehl und 'Gerichtsbarkeitserlass Barbarossa' in neuer Sicht," *VfZ* 25 (1977): 682–758.

32. For a discussion of the "necessitarian" argument in war, see M. Walzer, *Just and Unjust Wars*, 3d ed. (Harmondsworth, 1980), pp. 144–51 on the Laconia Affair; pp. 255–69 on the bombing of Germany and Japan; pp. 317–19 on the American bombing of St. Lô.

33. See *infra*, n. 6; Seidler, *Prostitution*, pp. 233–73, 312–17.

34. Apart from the literature already mentioned, see further on the criminal activities of the Wehrmacht in the East and their ideological background in C.

Streit, "The German Army and the Policies of Genocide," in *The Policies of Genocide*, ed. G. Hirschfeld (London, 1986), pp. 1–14; and ibid., J. Förster, "The German Army and the Ideological War against the Soviet Union," pp. 15–29; and idem., "New Wine in Old Skins? The Wehrmacht and the War of 'Weltanschauungen,' 1941," in *The German Military in the Age of Total War*, ed. W. Deist (Leamington Spa/New Hampshire, 1985), pp. 304–22.

35. J. Förster, "Das Unternehmen 'Barbarossa' als Eroberungs- und Vernichtungskrieg," in *Das deutsche Reich und der Zweite Weltkrieg*, 4:437–38; Streit, *Keine Kameraden*, p. 49; Krausnick, *Die Truppe*, pp. 121–26. Some historians were taken in by the generals' postwar falsifications. See, e.g., A. Dallin, *German Rule in Russia 1941–45*, 2d ed. (London, 1981), pp. 30–34. It is noteworthy that Dallin failed to correct this error even in the new edition of his work.

36. A different example of the function of structured chaos in ordering society is the Medieval carnival. See E. Le Roy Ladurie, *Carnival in Romans*, 3d ed. (Harmondsworth, 1981).

37. For a typical example of the lack of comprehension of the essential character of the war in the East by an otherwise highly sophisticated historian, see H. Mommsen, "Kriegserfahrungen," in *Über Leben im Krieg*, ed. U. Borsdorf and M. Jamin (Reinbeck bei Hamburg, 1989), p. 13. Even more bizarre is the fact that the relevant passage of this introduction entirely misrepresents the main argument of the contribution to the same volume on the Eastern Front, as well as the horrifying photographic evidence which accompanies it. See ibid., Bartov, *Extremfälle*; and "Bildserien: Verbrannte Erde; 'Partisanenbekämpfung;' eine Erschiessung sowjetischer Soldaten; sowjetische Juden als Zwangsarbeiter der Wehrmacht," pp. 162–68.

38. See also P. Hayes, *Industry and Ideology. IG Farben in the Nazi Era*, 2d ed. (Cambridge, 1989).

39. Bähr, *Kriegsbriefe*, p. 167.

40. It also turned the population against the occupiers, of course. This, along with Nazi ideological determinants, greatly hampered the army's attempt to establish an anti-Bolshevik Russian organization. On the most prominent collaborator, see C. Andreyev, *Vlasov and the Russian Liberation Movement. Soviet reality and émigré theories*, 2d ed. (Cambridge, 1989).

41. See, e.g., R. O. Paxton, *Vichy France*, 2d ed. (New York, 1975), pp. 292–93.

42. G. R. Ueberschär and W. Wette (eds.), *"Unternehmen Barbarossa"* (Paderborn, 1984), pp. 173–95 (contribution by R.-D. Müller); Müller, *Das Scheitern*. See also, D. Eichholz, "Der 'Generalplan Ost,'" *JfG* 26 (1982): 217–74; R.-D. Müller, "Industrielle Interessenpolitik im Rahmen des 'Generalplan Ost,'" *MGM* 1 (1981): 101–41; H. Heiber, "Der Generalplan Ost," *VfZ* (1958): 281–325; T. P. Mulligan, "Reckoning the Cost of People's War: the German Experience in the Central USSR," *RH* 5 (1982): 27–48; R.-D. Müller, *Das Tor zur Weltmacht* (Boppard am Rhein, 1984); Herbert, *Fremdarbeiter*; E. L. Homze, *Foreign Labour in Nazi Germany* (Princeton: N.J., 1967); N. Müller, "Dokumente zur Rolle der

Wehrmacht bei der Deportation sowjetischer Bürger zur Zwangsarbeit in Deutschland 1941–44," *BAZW* 4 (1970): 29–62; H. Pfahlmann, *Fremdarbeiter und Kriegsgefangene in der deutschen Kriegswirtschaft 1939–45* (Darmstadt, 1968).

43. BA-MA, RH27–18/4, 19.6.41.

44. BA-MA, RH27–18/153, 28.7.41, 10.8.41; BA-MA, RH27–18/26, 5.8.41; BA-MA, RH27–18/176, 6.8.41.

45. BA-MA, RH27–18/27, 17.8.41; BA-MA, RH27–18/181, reports for August, September, and November 1941; BA-MA, RH27–18/176, 7.11.41.

46. BA-MA, RH26–12/129, 26.5.–15.12.41; 11.8.41.

47. BA-MA, RH26–12/126, 7.11.41.

48. BA-MA, RH27–18/177, 7.12.41.

49. BA-MA, RH27–18/177, 8.11.–15.11.41.

50. BA-MA, RH27–18/78, 9.12.41.

51. BA-MA, RH27–18/24, 11.7.41.

52. BA-MA, RH27–18/24, 18.7.41; BA-MA, RH27–18/175, 27.7.41.

53. BA-MA, RH27–18/175, 30.7.41; BA-MA, RH27–18/26, 7.8.41.

54. BA-MA, RH27–18/27, 14.9.41.

55. BA-MA, RH26–12/128, 2.7.41; BA-MA, RH26–12/211, 17.8.41; BA-MA, RH26–12/294, 8.10.41.

56. BA-MA, RH27–18/181, 2.11.41.

57. BA-MA, RH26–12/126, 7.11.41.

58. BA-MA, RH26–12/291, 24.1.42.

59. BA-MA, RH26–12/288, 2.4.42.

60. BA-MA, RH27–18/75, 15.1.42; BA-MA, RH27–18/178, 28.2.42; BA-MA, RH27–18/76, 24.3.42; BA-MA, RH27–18/194, 9.11.42.

61. BA-MA, RH26–12/298, 29.6.43.

62. BA-MA, RH26–12/140, 16.12.41–31.8.42.

63. Guderian, *Panzer Leader*, p. 194.

64. BA-MA, RH27–18/184, reports for January, February, and March 1942.

65. BA-MA, RH27–18/76, 24.3.42; BA-MA, RH27–18/188, 17.4.42 and numerous other documents in the file, as well as in BA-MA, RH27–18/194 and BA-MA, RH27–18/196; BA-MA, RH27–18/178, 17.2.43.

66. Mebes, pp. 10–11.

67. BA-MA, RH26–1005/62, 1.9.–31.10.41.

68. BA-MA, RH26–1005/62, 16.9.42, 23.9.42; BA-MA, RH26–1005/37, 21.9.42.

69. BA-MA, RH26–1005/42, 15.10.42.

70. BA-MA, RH26–1005/71, 19.5.43.

71. BA-MA, RH26–12/246, 14.1.42; BA-MA, RH26–12/42, 16.1.42.

72. BA-MA, RH26–12/62, 27.2.42.

73. BA-MA, RH26–12/292, 25.6.42.

74. BA-MA, RH26–12/295, 11.10.42, 22.10.42, and 9.11.42 for an example of another "evacuation."

75. BA-MA, RH27–18/142, 24.7.43.

76. BA-MA, RH26–12/298, 1.7.43.

77. BA-MA, RH26–12/128, 21.9.41. See also Streit, *Keine Kameraden*, p. 140.

78. BA-MA, RH27–18/133, 26.3.43; BA-MA, RH27–18/199, 7.6.43; BA-MA, RH26–12/134, 5.8.42. On shortages of drugs and equipment, and the debate over possibly excessive amputations within the ranks of the Wehrmacht itself, see M. Kater, *Doctors under Hitler* (Chapel Hill: N.C., 1989), pp. 52–53. One of my readers (citing: *Heeres-Sanitäts-Inspekteur*, 17.10.42, microcopy T-78, reel 482, frame 6467197, National Archives) has drawn my attention to a Russian document claiming that the Germans killed their own litter bearers in order to prevent the spread of typhus. However, I have not found any confirmation for such allegations in the Wehrmacht documents I have seen.

79. BA-MA, RH26–12/292, 25.5.42; BA-MA, RH26–12/12, 12.6.42; BA-MA, RH26–12/137, reports between 8.7.–27.11.42; BA-MA, RH26–12/148, 17.3.43.

80. BA-MA, RH27–18/188, 14.5.42; BA-MA, RH27–18/189, 8.9.42.

81. BA-MA, RH26–12/211, 17.8.41; BA-MA, RH26–12/294, 22.4.42.

82. BA-MA, RH26–12/133, 17.9.42; BA-MA, RH26–12/295, 10.11.42.

83. BA-MA, RH27–18/179, 9.12.41.

84. Bähr, *Kriegsbriefe*, pp. 223–24.

85. BA-MA, RH27–18/74, 20.12.41, 21.12.41; BA-MA, RH27–18/78, 29.12.41; BA-MA, RH27–18/157, 1.1.42.

86. BA-MA, RH27–18/82, 28.1.42.

87. BA-MA, RH27–18/131, 13.2.43; BA-MA, RH27–18/203, 3.8.43.

88. BA-MA, RH26–12/133, 12.2.43; BA-MA, RH26–12/72, 14.2.43, 15.2.43.

89. BA-MA, RH26–12/141, 28.2.43; BA-MA, RH26–12/76, 1.3.43.

90. BA-MA, RH26–1005/68, 15.1.–27.2.43; BA-MA, RH26–1005/71, 2.8.43; BA-MA, RH26–1005/70, 1.4.–30.9.43.

91. BA-MA, RH26–1005/71, 6.9.–28.9.43; Benes, pp. 22–23, entry for 17.9.43.

92. Ueberschär, *Barbarossa* (contribution by C. Streit), pp. 197–218. See further in Streit, *Keine Kameraden*; A. Streim, *Die Behandlung sowjetischer Kriegsgefangener in "Fall Barbarossa"* (Heidelberg, 1981); M. Messerschmidt, "Kommandobefehl und NS-Völkerrechtsdenken," *RDP* 11 (1972): 110–34. More generally on POWs, see G. H. Davis, "Prisoners of War in Twentieth-Century War Economies," *JCH* 12 (1977): 623–34; A. Rosas, *The Legal Status of Prisoners of War* (Helsinki, 1976).

93. BA-MA, RH26–12/244, 22.6.41.

94. BA-MA, RH26–12/85, 21.5.42.

95. BA-MA, RH26–12/243, 13.9.41.

96. BA-MA, RH26–12/56.

97. BA-MA, RH26–12/27, 5.7.41.

98. BA-MA, RH26–12/27, 17.7.41.

99. BA-MA, RH26–12/24, 25.6.41.

100. BA-MA, RH26–12/175, 30.6.41.

101. Guderian, *Panzer Leader*, p. 152.

102. BA-MA, RH27–18/157, 14.2.42.

103. Bartov, *Eastern Front,* pp. 75–76.

104. BA-MA, RH26–1005/42, 19.9.42.

105. BA-MA, RH26–1005/42, 23.9.42.

106. BA-MA, RH26–1005/47, 26.4.43.

107. BA-MA, RH26–12/211, 31.7.41.

108. BA-MA, RH27–18/27, 17.8.41.

109. BA-MA, RH26–12/288, 6.8.41; BA-MA, RH26–12/290, 21.10.41; Streit, *Keine Kameraden,* pp. 138–40 for calorie tables.

110. BA-MA, RH27–18/175, 17.6.41.

111. BA-MA, RH26–12/291, 24.1.42; BA-MA, RH26–12/246, 28.1.42; BA-MA, RH27–18/194, 25.9.42.

112. BA-MA, RH26–12/287, 21.8.41; BA-MA, RH27–18/26, 6.8.41.

113. BA-MA, RH26–12/295, 14.11.41; BA-MA, RH26–12/51, 6.3.42; BA-MA, RH26–12/220, 8.12.42, 14.12.42; BA-MA, RH26–12/59, 1.2.43; BA-MA, RH26–12/78, 9.3.43, 1.4.43, 1.5.43, 1.6.43; BA-MA, RH27–18/76, 2.11.41, 7.3.42; BA-MA, RH27–18/177, 8.-15.11.41; BA-MA, RH27–18/73, 11.11.41; BA-MA, RH27–18/188, 2.5.42; BA-MA, RH27–18/189, 25.6.42; BA-MA, RH27–18/124, 26.7.42, 25.11.42, 6.12.42, 15.12.42; BA-MA, RH27–18/194, 19.8.42; BA-MA, RH27–18/159, 5.9.42; BA-MA, RH27–18/164, 1.4.–30.6.43; BA-MA, RH27–18/196, 30.4.43; BA-MA, RH27–18/144, 24.8.43.

114. BA-MA, RH26–12/244, 22.6.41.

115. BA-MA, RH26–12/211, 31.7.41.

116. BA-MA, RH27–18/26, 4.8.41.

117. BA-MA, RH27–18/28, 14.8.41; BA-MA, RH27–18/242, 25.8.41; BA-MA, RH27–18/154, September 1941.

118. BA-MA, RH26–12/32, 5.8.41.

119. See esp. Streit, *Keine Kameraden,* pp. 28–61; and the detailed examination in Krausnick, *Die Truppe.*

120. BA-MA, RH26–12/126, 4.7.41.

121. See, e.g., BA-MA, RH26–12/128, 12.8.41.

122. BA-MA, RH26–12/287, 15.8.41.

123. BA-MA, RH26–12/242, 25.8.41; BA-MA, RH26–12/154, September 1941; BA-MA, RH26–12/123, 18.9.41.

124. BA-MA, RH27–18/154, 8.10.41.

125. BA-MA, RH26–12/128, 27.9.41, 11.10.41; BA-MA, RH26–12/243, 4.10.41.

126. BA-MA, RH26–12/245, 13.10.41.

127. BA-MA, RH26–12/244, 31.10.41, 14.11.41; BA-MA, RH26–12/206, 1.11.41.

128. BA-MA, RH26–12/245, 17.11.41, 20.11.41, 2.12.41, 11.12.41.

129. BA-MA, RH26–12/200, 5.12.41; BA-MA, RH26–12/246, 5.12.41.

130. BA-MA, RH26–12/246, 29.11.–5.12.41.

131. BA-MA, RH26–12/245, 3.12.41, 7.12.41.

132. BA-MA, RH26–12/245, 14.12.41; BA-MA, RH26–12/294, 9.1.42.

133. BA-MA, RH26–12/246, 19.12.41, 24.12.41.

134. BA-MA, RH26–12/246, 4.1.42, 8.1.42.

135. BA-MA, RH26–12/246, 31.1.42.

136. See, e.g., the destruction of the village Nov. Ladomiry: BA-MA, RH26–12/84, 20.1.–30.1.42.

137. See, e.g., BA-MA, RH27–18/157, 11.2.42; BA-MA, RH27–18/158, 7.4.42–3.2.43, esp. entries for 7.4.–1.7.42 on evacuations; BA-MA, RH27–18/115, 13.4.42.

138. BA-MA, RH26–1005/6, 15.4.42.

139. BA-MA, RH26–1005/42, 23.9.42.

140. BA-MA, RH27–18/162, 4.2.-2.6.43, esp. entries for 12.5.–16.5.43; BA-MA, RH27–18/132, 10.5.43, 9.6.43; BA-MA, RH27–18/164, 13.5.43, and *Anlage 426*; BA-MA, RH27–18/196, 16.5.43; BA-MA, RH27–18/130, 16.5.43; BA-MA, RH27–18/135, 2.6.43; BA-MA, RH27–18/166, 8.6.43. Also see Krausnick, *Die Truppe*, pp. 243–49.

141. BA-MA, RH26–12/244, 11.7.41.

142. BA-MA, RH26–12/245, 13.10.41.

143. BA-MA, RH27–18/24, 24.6.41, 8.7.41.

144. BA-MA, RH27–18/159, 8.6.42.

145. See a good example of this practice in C. R. Browning, *Fateful Months. Essays on the Emergence of the Final Solution* (New York, 1985), pp. 39–56.

146. BA-MA, RH26–12/245, 13.10.41.

147. BA-MA, RH26–12/246, 11.1.42. Also see, e.g., BA-MA, RH26–12/83, 16.12.41–28.2.43. On the killing of Jewish children by the Wehrmacht, see, e.g., H. Krausnick and H. C. Deutsch (eds.), *Helmut Groscurth: Tagebücher eines Abwehroffiziers* (Stuttgart, 1970), pp. 234–42.

148. BA-MA, RH26–1005/42, 22.10.42.

149. BA-MA, RH26–12/245, 13.10.41.

150. BA-MA, RH27–18/177, 17.11.41; BA-MA, RH27–18/164, 30.4.43.

151. BA-MA, RH26–1005/42, 22.10.42.

152. M. Messerschmidt, "Deutsche Militärgerichtsbarkeit im Zweiten Weltkrieg," in *Recht, Verwaltung und Justiz im Nationalsozialismus*, ed. M. Hirsch et al. (Köln, 1984), pp. 553–56; idem., *German Military Law*; and most recently, idem. and F. Wüllner, *Die Wehrmachtjustiz im Dienste des Nationalsozialismus. Zerstörung einer Legende* (Baden-Baden, 1987). Also see O. P. Schweling, *Die deutsche Militärjustiz in der Zeit des Nationalsozialismus*, 2d ed. (Marburg, 1978), pp. 216–84; Seidler, *prostitution*, pp. 205–6, 271. For a touching account of the execution of a Wehrmacht deserter in Holland, see letter by Friedrich Andreas von Koch, dated 3 October 1942, cited in Bähr, *Kriegsbriefe*, pp. 238–40; and for the case of Ernst Jünger, son of the author, who shortly before he was killed in Italy at the age of 18 was arrested for allegedly criticizing Hitler, see his letters, cited ibid., pp. 394–97; also see his father's war diary, E. Jünger, *Strahlungen II*, in *Tagebücher III*, vol. 3 of *Sämtliche Werke* (Stuttgart, 1979 [1949]), pp. 231, 249–52, 322, 337–38, 344, 360–63.

153. BA-MA, RH27–18/4, 19.6.41.

154. BA-MA, RH27–18/176, 1.8.41.

155. BA-MA, RH27–18/28, 18.8.41.

156. BA-MA, RH27–18/74, 12.12.41.

157. BA-MA, RH27–18/74, 30.12.41.

158. BA-MA, RH27–18/73, 29.11.41.

159. BA-MA, RH27–18/76, 12.3.42.

160. BA-MA, RH27–18/76, 19.3.42.

161. BA-MA, RH27–18/123, 27.9.42, 28.9.42.

162. BA-MA, RH26–1005/61, 7.4.43; BA-MA, RH26–1005/70, *KTB Ib Nr. 3*, 1.4.–30.9.43, p. 102, entry for 4.-8.8.43.

163. BA-MA, RH26–12/131, 25.12.41; BA-MA, RH26–12/45, 5.10.41; BA-MA, RH26–12/139, 4.5.43; BA-MA, RH26–12/151, 24.9.43.

164. BA-MA, RH26–12/85, 24.10.42.

165. BA-MA, RH26–12/267, 7.5.42.

166. BA-MA, RH26–12/85, 27.5.42.

167. BA-MA, RH26–12/134, 30.7.42.

168. BA-MA, RH27–18/131, 25.2.43.

169. BA-MA, RH26–12/72, 15.2.43.

170. BA-MA, RH27–18/142, 17.7.43.

171. BA-MA, RH27–18/24, 26.6.41.

172. BA-MA, RH27–18/24, 4.7.41.

173. BA-MA, RH27–18/147, 12.7.43.

174. Paul, *18.Pz.Div.*, p. 262.

175. BA-MA, RH27–18/142, 17.7.43.

176. BA-MA, RH27–18/142, 19.7.43.

177. Spaeter/Schramm, *GD*, 3: 665–81.

178. Klee, *Schöne Zeiten*, pp. 122–29; Borsdorf, *Über Leben*, p. 165.

Chapter 4

1. Buchbender/Stertz, *Das andere Gesicht*, p. 74, letter 104.

2. Generally on indoctrination and propaganda in the Third Reich, see J. W. Baird, *The Mythical World of Nazi War Propaganda, 1939–45* (Minneapolis: Minnesota, 1974); M. Balfour, *Propaganda in War, 1939–45* (London, 1979); E. K. Bramsted, *Goebbels and National Socialist Propaganda, 1925–45* (London, 1965); R. Cecil, *The Myth of the Master Race* (London, 1972); Messerschmidt, *Die Wehrmacht*; O'Neill, *The German Army*; J. Sywottek, *Mobilmachung für den Krieg* (Opladen, 1976); D. Welch (ed.), *Nazi Propaganda* (London, 1983); Z. A. B. Zeman, *Nazi Propaganda* (London, 1964).

3. On Hitler's ideology and its implementation, see E. Jäckel, *Hitlers Weltanschauung: Entwurf einer Herrschaft*, 3d ed. (Stuttgart, 1986); and idem., *Hitlers Herrschaft: Vollzug einer Weltanschauung* (Stuttgart, 1986).

4. For an emphasis on the "organization of power" as a central component in the army's collaboration with Hitler, see M. Geyer, "Etudes in Political History:

Reichswehr, NSDAP, and the Seizure of Power," in *The Nazi Machtergreifung*, ed. P. D. Stachura (London, 1983), pp. 101–23; also idem., "Professionals and Junkers: German Rearmament and Politics in the Weimar Republic," in *Social Change and Political Development in Weimar Germany*, ed. R. Bessel and E. J. Feuchtwanger (London, 1981), pp. 77–133; and idem., *Aufrüstung oder Sicherheit. Die Reichswehr in der Krise der Machtpolitik 1924–36* (Wiesbaden, 1980). On institutional and personal aspects, see Müller, *The Army*. On ideological aspects, see B. Hüppauf, "Langemarck, Verdun and the Myth of a New Man in Germany after the First World War," *W&S* 6 (1988): 70–103.

5. Powerful examples of this process can be found throughout Klee, *Schöne Zeiten*.

6. On education see H. Scholtz, *Erziehung und Unterricht unterm Hakenkreuz* (Göttingen, 1985); G. Planter (ed.), *Schule im Dritten Reich—Erziehung zum Tod?*, 2d ed. (München, 1984); K.-I. Flessau et al. (eds.), *Erziehung im Nationalsozialismus* (Köln, 1987); G. W. Blackburn, *Education in the Third Reich* (New York, 1959); E. Nyssen, *Schule im Nationalsozialismus* (Heidelberg, 1979). On the universities, see R. G. S. Weber, *The German Student Corps in the Third Reich* (London, 1986); G. J. Giles, *Students and National Socialism in Germany* (Princeton: N.J., 1985); J. R. Pauwels, *Women, Nazis, and Universities* (Westport: Conn., 1984); M. Burleigh, *Germany turns Eastwards. A Study of Ostforschung in the Third Reich*, 2d ed. (Cambridge, 1989). On the Hitler Youth, see A. Klönne, *Jugend im Dritten Reich*, 2d ed. (Köln, 1984); H. W. Koch, *The Hitler Youth* (London, 1975); P. D. Stachura, *Nazi Youth in the Weimar Republic* (Santa Barbara: Calif., 1975); F. J. Stephens, *Hitler Youth* (London, 1973). An excellent contemporary account in E. Mann, *School for Barbarians* (London, 1939). For some illuminating general surveys, see F. Ringer, "Bildung, Wirtschaft und Gesellschaft in Deutschland 1800–1960," *GuG* 6 (1980): 5–35; H. Kupffer, *Der Faschismus und das Menschenbild der deutschen Pädagogik* (Frankfurt/M., 1984); J. M. Mushaben, "Youth Protest and the Democratic State: Reflections on the Rise of Anti-Political Culture in Prewar Germany and the German Federal Republic," *RPS* 2 (1986): 171–97. On the impact of Nazi propaganda in the 1920s, see I. Kershaw, "Ideology, Propaganda, and the Rise of the Nazi Party," in *The Nazi Machtergreifung*, ed. P. D. Stachura (London, 1983), pp. 162–81.

7. In this context see also A. Kaes, *From Hitler to Heimat: the Return of History as Film* (Cambridge: Mass., 1989).

8. A. Heck, *A Child of Hitler*, 3d ed. (Toronto/New York/London, 1986).

9. On the cinema in the Third Reich, see D. S. Hull, *Film in the Third Reich* (Berkeley: Calif., 1969); D. Welch, "Nazi Wartime Newsreel Propaganda," in *Film & Radio Propaganda in World War II*, ed. K. R. M. Short (Knoxville: Tenn., 1983), pp. 201–19.

10. Cited in R. Schörken, "Jugendalltag im Dritten Reich," in *Geschichte im Alltag—Alltag in der Geschichte*, ed. K. Bergmann and R. Schörken (Düsseldorf, 1982), pp. 238–39. This excellent article analyzes a number of autobiographies whose authors were children or young men in the Third Reich; I have drawn on its insights for this and a few more examples below.

11. Ibid., pp. 239–40. See further on the indoctrination of the youth in the Third Reich, in N. A. Huebsch, "The 'Wolf Cubs' of the New Order: The Indoctrination and Training of the Hitler Youth," in *Nazism and the Common Man*, ed. O. C. Mitchell (Washington D. C., 1981), pp. 93–114; D. J. K. Peukert, *Inside Nazi Germany* (London, 1987), pp. 145–74; P. Loewenberg, "The Psychological Origins of the Nazi Youth Combat," *AHR* 76 (1971): 1457–1502.

12. Schörken, *Jugendalltag*, pp. 240–41.

13. Cited in L. Niethammer, "Heimat und Front. Versuch, zehn Kriegserinnerungen aus der Arbeiterklasse des Ruhrgebietes zu verstehen," in *Die Jahre weiss man nicht, wo man die heute hinsetzen soll. Faschismus im Ruhrgebiet,* ed. L. Niethammer (Berlin, 1983), pp. 209–13. This is a careful and highly important analysis of oral testimonies of workers who had served in the Wehrmacht. On the issue of workers as soldiers, also see O. Bartov, "The Missing Years: German Workers, German Soldiers," *GH* 8 (1990): 46–65. On workers and the *Volksgemeinschaft*, see N. Frei, *Der Führerstaat. Nationalsozialistische Herrschaft 1933 bis 1945* (München, 1987), pp. 93–100.

14. Niethammer, *Heimat und Front,* pp. 213–18. Further on the relationship between the workers and the regime, see M. Ruck, *Bollwerk gegen Hitler? Arbeiterschaft, Arbeiterbewegung und die Anfänge des Nationalsozialismus* (Köln, 1988); D. J. K. Peukert and F. Bajohr, *Spuren des Widerstands: Die Bergarbeiterbewegung im Dritten Reich und im Exil* (München, 1987); A. Merson, *Communist Resistance in Nazi Germany* (London, 1987).

15. H. Böll, *Was soll aus dem Jungen bloss werden?* (Bornheim, 1981).

16. H. Böll, *Der Zug war pünktlich,* 12th ed. (München, 1980).

17. B. Engelmann, *In Hitler's Germany. Daily Life in the Third Reich* (New York, 1986), p. xii.

18. Ibid., pp. 27, 34.

19. Ibid., p. 117.

20. Ibid., pp. 8–9.

21. Schörken, *Jugendalltag*, pp. 242–44.

22. BA-MA, RH27–18/27, 3.8.41.

23. Richardson, *Sieg Heil!,* p. 1.

24. Ibid., p. 3.

25. Ibid., pp. 36–37. Nor were the parents always passive observers. Frau Anna Neuber, for instance, recounts that when told by her son Georg of the atrocities he had witnessed in Poland, she urged him "always to obey his superiors, no matter what, and to pray to God that he would never be forced to do anything that was wrong." Engelmann, *Hitler's Germany,* p. 174.

26. Richardson, *Sieg Heil!,* pp. 104–5.

27. Ibid., 108–9.

28. Bähr, *Kriegsbriefe,* p. 160.

29. Ibid., p. 161.

30. Ibid., pp. 106–7.

31. Ibid., p. 111.

32. Ibid., p. 113.

33. G. Sajer, *The Forgotten Soldier*, 2d ed. (London, 1977), pp. 263–67.

34. Text in Wheeler-Bennett, *Nemesis of Power*, p. 339.

35. Apart from the literature cited *infra*, n. 2, also see Bartov, *Eastern Front*, pp. 68–105, which argues that the indoctrination material disseminated by the Wehrmacht reached even the remotest units at the front and that it was by and large welcomed by junior officers and soldiers alike. A recent book argues for a different view, though it ultimately does not reach radically contradictory conclusions: Schulte, *The German Army*.

36. See, e.g., J. Fest, *Hitler*, 3d ed. (Harmondsworth, 1982), pp. 521–22.

37. F. Stern, *Dreams and Delusions: The Drama of German History* (London, 1988), p. 153. It has similarly been claimed that during the First World War "the figure of Hindenburg, with clever propaganda manipulations, and by the fact that he incorporated the hopes and aspirations of so many Germans, had taken on mythical proportions," and that "[b]y some curious workings of social psychology it seems that Hindenburg and Ludendorff appeared more infallible and invincible the more it became apparent that Germany's military position was precarious in the extreme." M. Kitchen, *The Silent Dictatorship. The Politics of the German High Command under Hindenburg and Ludendorff, 1916–1918* (New York, 1976), pp. 273, 277.

38. U. Tal, *"Political Faith" of Nazism Prior to the Holocaust* (Tel-Aviv, 1978), p. 7.

39. Engelmann, *Hitler's Germany*, pp. 38–39.

40. Ibid., p. 189.

41. Tal, *"Political Faith,"* p. 9.

42. Ibid., p. 30.

43. On the related issue of racial hygiene, the pre-Nazi roots of the idea of creating a "healthy" community and doing away with inferior elements, as well as on the implementation of this ideology in the Nazi euthanasia campaign, see, e.g., H. W. Schmuhl, *Rassenhygiene, Nationalsozialismus, Euthanasie* (Göttingen, 1987); C. Ganssmüller, *Die Erbgesundheitspolitik des Dritten Reiches* (Köln and Wien, 1987).

44. BA-MA, RH26–12/227, *"Der Führer sprach am Heldengedenktag 1940,"* 6.4.1940.

45. BA-MA, RH26–12/227, *"Rundfunkvortrag Nr. 12: 'Die Stimme der Soldaten,' "* 25.4.40.

46. I. Kershaw, *The "Hitler Myth." Image and Reality in the Third Reich* (Oxford, 1987), pp. 151–68.

47. BA-MA, RH26–12/227, *"Mitteilungen für die Truppe, Nr. 10, Juni 1940."*

48. BA-MA, RH26–12/294, *"Mitteilungen für das Offizierkorps, Nr. 4 April 1942."*

49. BA-MA, RH26–12/298, *"Nationalpolitischer Unterricht im Heere, für den Kompanie-Führer, Nr. 26/Februar 1943."*

50. Klee, *Schöne Zeiten*, p. 49.

51. *"Mitteilungen für die Truppe, Nr. 116,"* cited in Messerschmidt, *Die Wehrmacht*, pp. 326–27.

52. See Kershaw, *Popular Opinion*, p. 360; S. Gordon, *Hitler, Germans, and the "Jewish Question"* (Princeton: N.J., 1984), pp. 185–86.

53. This theme is most clearly and succinctly examined in Müller, *The Army*, pp. 16–53.

54. Letter dated 11 December 1938, cited in Wheeler-Bennett, *Nemesis of Power*, p. 380.

55. Ibid., citing von Hassel's diary entry for 18 December 1938. See also Ludwig Beck's letter of November 1918, cited in Müller, *Beck*, pp. 323–28.

56. Förster, *Hitlers Entscheidung*, p. 24.

57. Förster, *Unternehmen "Barbarossa,"* p. 446.

58. BA-MA, RH27–18/4, 21.6.41.

59. Cited in Streit, *Keine Kameraden*, p. 115.

60. Ibid., p. 116. On plans made by the German army in the First World War according to which the Jews were to be removed [*verpflanzt*] from a wide frontier strip between Poland and the Reich, see Kitchen, *Silent Dictatorship*, pp. 142, 193–94.

61. Streit, *Keine Kameraden*, pp. 116–17.

62. BA-MA, RH26–12/262, 28.12.41.

63. BA-MA, RH26–12/297, 27.1.43.

64. Förster, *Hitlers Entscheidung*, p. 25.

65. BA-MA, RH26–12/16, 8.7.40.

66. BA-MA, RH26–12/298, 6.6.43.

67. BA-MA, RH26–12/89, 18.6.43.

68. BA-MA, RH27–18/159, 27.10.42.

69. BA-MA, RH26–1005/7, 3.4.-3.7.42.

70. BA-MA, RH26–1005/39, 22.9.42.

71. BA-MA, RH26–1005/47, 10.4.43.

72. BA-MA, RH26–1005/47, 11.5.43.

73. *"Mitteilungen für die Truppe, Nr. 389, Januar 1945"* reprinted this "credo." Cited in Messerschmidt, *Die Wehrmacht*, pp. 331–32.

74. V. R. Berghahn, "NSDAP und 'Geistige Führung' der Wehrmacht," *VfZ* 17 (1969): 17–71; G. L. Weinberg, "Dokumentation: Adolf Hitler und der NS-Führungsoffizier (NSFO)," *VfZ* 12 (1964): 443–56; Messerschmidt, *Die Wehrmacht*, pp. 441–80; R. L. Quinnett, "The National Socialist Leadership Officers," (Univ. of Oklahoma Ph.D. thesis, 1973).

75. Cite in Messerschmidt, *Die Wehrmacht*, p. 466.

76. See above, Chapter 3, for Wehrmacht instructions in the West to execute German emigrants, that is, citizens who had fled either for political or "racial" reasons from the Reich. But see R. Hilberg, *The Destruction of European Jews*, rev. ed. (New York, 1986), 2:627, who writes that non-German Jewish POWs generally "enjoyed relative immunity." And D. A. Foy, *For You the War is Over. American Prisoners of War in Nazi Germany* (New York, 1984), p. 128, according to whom "[t]here are as many examples of humane treatment of American black and Jewish POWs as there are of harsh treatment."

77. On the so-called *"Traditionsdebatte"* regarding the Bundeswehr, see D. Abenheim, *Reforging the Iron Cross. The Search for Tradition in the West German Armed Forces* (Princeton: N.J., 1988).

78. See former American President Ronald Reagan's speech about his visit to the German military and SS cemetery in Bitburg: "I think that there's nothing wrong with visiting that cemetery where those young men are victims of Nazism also, even though they were fighting in the German uniform, drafted into service to carry out the hateful wishes of the Nazis." Cited in *Time Magazine*, 29 April 1985, No. 17, p. 44.

79. Guderian, *Panzer Leader*, p. 440.

80. Ibid., p. 436.

81. Ibid., p. 462.

82. B. H. Liddell Hart, *The Other Side of the Hill* (London, 1948), p. 29.

83. D. Brader, "Foreword" to H.-U. Rudel, *Stuka Pilot*, 2d ed. (Maidstone, 1973).

84. Ibid., p. 189.

85. For some perceptive analyses of the so-called *Historikerstreit*, see C. S. Maier, *The Unmasterable Past* (Cambridge: Mass., 1988); H.-U. Wehler, *Entsorgung der deutschen Vergangenheit?* (München, 1988); I. Kershaw, *The Nazi Dictatorship*, 2d ed. (London, 1989), pp. 168–91; and idem., "Neue deutsche Unruhe? Das Ausland und der Streit um die deutsche National- und Zeitgeschichte," in *Streitfall Deutsche Geschichte*, ed. Landeszentrale für politische Bildung Nordrhein-Westfalen (n.d.), pp. 111–30; R. J. Evans, *In Hitler's Shadow* (London, 1989); S. Friedländer, "Some Reflections on the Historisation of National Socialism," *TAJB* 16 (1987): 310–24; P. Baldwin (ed.), *Reworking the Past* (Boston, 1990).

86. The other major contributions to the "revisionist" approach are encapsulated in: M. Stürmer, "Geschichte in geschichtslosem Land," *FAZ*, 25.4.86; E. Nolte, "Vergangenheit, die nicht vergehen will," *FAZ*, 6.6.86; but see also idem., "Between Myth and Revisionism?," in *Aspects of the Third Reich*, ed. W. Koch (London, 1985), pp. 17–38. For a more complex "revisionist" position, see Broszat, *Plädoyer*. The first and most devastating, though not unproblematic attack on this approach is J. Habermas, "Eine Art Schadensabwicklung," *Die Zeit*, 11.7.86. The above cited newspaper articles, and many other important contributions to this debate, are reprinted in *"Historikerstreit."*

87. Hillgruber, *Zweierlei Untergang*, pp. 24–25.

88. Ibid., pp. 34–35.

89. Ibid., p. 53.

90. Ibid., p. 47.

91. Generally on civilian and military conservative opposition, see, inter al., H. C. Deutsch, *The Conspiracy against Hitler in the Twilight War* (Minneapolis: Minn., 1968), and idem., *Hitler and his Generals* (Minneapolis: Minn., 1974); P. Hoffmann, *German Resistance to Hitler* (Cambridge: Mass., 1988); G. Ritter, *Carl Goerdeler und die deutsche Widerstands-Bewegung* (Stuttgart, 1954). And with mention also of left-wing opposition: H. Bull (ed.), *The Challenge of the Third*

Reich (Oxford, 1986); H. Graml et al., *The German Resistance to Hitler* (London, 1970).

92. Hillgruber, *Zweierlei Untergang,* p. 64.

93. Ibid.

94. For a more thorough treatment of Hillgruber's book, see Bartov, *Historians on the Eastern Front.*

95. Further on the problems and perils of writing the *Alltagsgeschichte* of the war, see O. Bartov, "Von unten betrachtet: Überleben, Zusammenhalt und Brutalität an der Ostfront," in *Zwei Wege nach Moskau. Vom Hitler-Stalin-Pakt zum "Unternehmen Barbarossa" 1939–1941,* ed. B. Wegner (München/Zürich, forthcoming March 1991).

96. M. G. Steinert, *Hitler's War and the Germans* (Athens: Ohio, 1977), p. 196. See also V. R. Berghahn, "Meinungsforschung im 'Dritten Reich,' " *MGM* 1 (1967): 83–119.

97. Kershaw, *Hitler Myth,* p. 209.

98. Steinert, *Hitler's War,* p. 272.

99. Ibid., pp. 282–83.

100. Ibid., p. 289.

101. *The Goebbels Diaries 1939–1941,* trans. and ed. F. Taylor, 2d ed. (Harmondsworth, 1984), p. 446.

102. Ibid., pp. 342–43.

103. *The Goebbels Diaries: The Last Days,* ed. H. Trevor-Roper, 2d ed. (London and Sydney, 1979), p. 21.

104. Ibid., p. 75.

105. Ibid., p. 80.

106. Ibid., p. 82.

107. Ibid., pp. 89, 95.

108. E. von Manstein, *Aus einem Soldatenleben* (Bonn, 1958), pp. 353–54.

109. Apart from this argument, Wehrmacht generals also claimed that they could not rebel because, as Manstein put it, their whole tradition was based on a "selfless devotion to the service of the state, Reich and Volk." Ibid. Similar arguments in the following: idem., *Verlorene Siege,* 2d ed. (Frankfurt/M., 1964), pp. 7–8; K. Doenitz, *Memoirs,* 2d ed. (London, 1959), pp. 299–314; Guderian, *Panzer Leader,* pp. 458–64; Kesselring, *The Memoirs* (London, 1953), pp. 314–15; S. Westphal, *The German Army in the West* (London, 1951), pp. 3–18; and many others.

110. J. von Herwarth, *Against Two Evils* (London, 1981), pp. 203 ff. This author records that his soldiers found it difficult to believe that the enemy was composed of human beings rather than *Untermenschen.* Many other officers noted the same reaction. See, e.g., H. Meier-Welcker, *Aufzeichnungen eines Generalstabsoffiziers 1939–42* (Freiburg, 1982), pp. 128–29, 148, 150, etc.

111. Herwarth, *Two Evils,* p. 255.

112. Ibid., p. 254.

113. Mommsen, *Kriegserfahrungen,* p. 13.

114. See esp. Klemperer, *Unbewältigte Sprache.*

115. Bähr, *Kriegsbriefe*, pp. 156–57.

116. Richardson, *Sieg Heil!*, p. 50.

117. BA-MA, RH26–12/238, 26.11.40, p. 7.

118. Ibid.

119. Bähr, *Kriegsbriefe*, p. 158.

120. Richardson, *Sieg Heil!*, p. 69. In this context, see also R. Cobb, *French and Germans, Germans and French. A Personal Account of France under Two Occupations 1914–1918/1940–1944*, 2d ed. (Hanover: N.H., 1984).

121. BA-MA, RH26–12/195, 6.1.41, pp. 5–6.

122. Ibid., p. 6.

123. Ibid.

124. Richardson, *Sieg Heil!*, pp. 72–73.

125. Ibid., p. 76.

126. Ibid., p. 78.

127. Ibid., p. 80.

128. Ibid., p. 81.

129. Bähr, *Kriegsbriefe*, pp. 51–52.

130. Buchbender/Stertz, *Das andere Gesicht*, p. 71, letter 96.

131. Ibid., pp. 71–72, letter 98.

132. Richardson, *Sieg Heil!*, p. 115.

133. Buchbender/Stertz, *Das andere Gesicht*, pp. 72–73, letter 101.

134. Ibid., p. 170, letter 345. In this context my attention has been drawn to the forthcoming essay by G. Cocks, "Partners and Pariahs: Jews and Medicine in Modern German Society," *LBIY* 36 (1991), concerning the association of Jews with disease also among civilians as a result of Eastern imperialism and enslavement.

135. Buchbender/Stertz, *Das andere Gesicht*, p. 170, letter 346.

136. Ibid., p. 76, letter 108.

137. Richardson, *Sieg Heil!*, p. 116.

138. Ibid., p. 119.

139. Ibid., p. 124.

140. Ibid., p. 132.

141. Buchbender/Stertz, *Das andere Gesicht*, p. 77, letter 111.

142. Ibid., p. 78, letter 115.

143. See, e.g., E. Topitsch, *Stalin's War* (New York, 1987). This is also the main argument of Hoffmann, *Die Sowjetunion*; and idem., *Die Kriegführung*.

144. Not in the sense meant in Nolte, *Myth and Revisionism*, but rather as a legitimation for barbarism. Nolte, however, has further pursued his idea. See E. Nolte, *Das Vergehen der Vergangenheit* (Berlin, Frankfurt/M., 1987).

145. *Deutsche Soldaten sehen die Sowjetunion* (Berlin, 1941), p. 49. Though this volume was published in the Third Reich with the obvious sanction of Goebbels, I see no reason to doubt the authenticity of the letters it contains, especially as they resemble soldiers' correspondence collected after the war. To be sure, the selection of the letters was done with an eye to their propagandistic value (which incidentally also reveals the knowledge among German soldiers of murders of Jews and "Bolsheviks" and their participation in such actions). Similarly, Bähr's *Kriegsbriefe* compiled let-

ters obviously intended to illustrate the students' piety and traditional romantic patriotism (using as a model the volume of war letters by students of the Great War, often referred to in the correspondence itself); and Buchbender's *Das andere Gesicht* presents letters with a political/ideological content. Yet I would claim that by putting all those collections together, along with documentary evidence such as censorship reports, and personal collections as Karl Fuchs', one does gain considerable insight particularly into the younger combat soldiers' perception of reality and ideological make-up, as well as into the vocabulary and imagery with which they expressed themselves and articulated their thoughts.

146. Ibid., p. 54.

147. Buchbender/Stertz, *Das andere Gesicht*, p. 78, letter 116.

148. *Deutsche Soldaten*, p. 60.

149. Ibid., p. 15.

150. Ibid., p. 19.

151. Ibid., p. 24.

152. Buchbender/Stertz, *Das andere Gesicht*, pp. 86–87, letter 143.

153. Richardson, *Sieg Heil!*, p. 157.

154. Buchbender/Stertz, *Das andere Gesicht*, p. 79, letter 118.

155. Richardson, *Sieg Heil!*, pp. 125–26.

156. Buchbender/Stertz, *Das andere Gesicht*, letter 127.

157. *Deutsche Soldaten*, pp. 12–14.

158. Ibid., p. 49.

159. Richardson, *Sieg Heil!*, pp. 144–45.

160. Ibid., p. 145.

161. Buchbender/Stertz, *Das andere Gesicht*, p. 80, letter 123.

162. *Deutsche Soldaten*, pp. 52–53.

163. Buchbender/Stertz, *Das andere Gesicht*, pp. 84–85, letter 137.

164. Ibid., p. 84, letter 135.

165. Ibid., p. 85, letter 139.

166. *Deutsche Soldaten*, pp. 17–18.

167. Richardson, *Sieg Heil!*, p. 122.

168. Ibid., p. 123.

169. Ibid., p. 138.

170. Ibid., p. 135.

171. Ibid., p. 147.

172. Buchbender/Stertz, *Das andere Gesicht*, p. 87, letter 144.

173. Bähr, *Kriegsbriefe*, p. 112.

174. *Deutsche Soldaten*, p. 35.

175. Ibid., p. 16.

176. Ibid., pp. 37–38. See also Klee, *Schöne Zeiten*, pp. 31–51 for eyewitness accounts and photographs of the pogroms in Lithuania; and Krausnick/Wilhelm, *Die Truppe*, for Army-*Einsatzgruppe* collaboration in the Kovno pogroms.

177. *Deutsche Soldaten*, p. 38.

178. Ibid., pp. 41–43.

179. Ibid., p. 44.

180. Ibid., pp. 44–45.

181. Ibid., p. 45.

182. Ibid., pp. 59–60.

183. Ibid., p. 43.

184. Ibid., pp. 43–44.

185. Buchbender/Stertz, *Das andere Gesicht*, p. 171, letter 349.

186. Ibid., letter 350.

187. This was also the policy in Serbia, the consequence of which was the destruction by the Wehrmacht of the entire Jewish male population there. See Browning, *Fateful Months*, pp. 39–56.

188. Buchbender/Stertz, *Das andere Gesicht*, letter 351.

189. Ibid., p. 172, letter 352.

190. Ibid., pp. 172–73, letter 353.

191. *Deutsche Soldaten*, p. 19.

192. Ibid., p. 40.

193. Ibid., p. 55.

194. All these citations are taken from an unpublished article by Mark Mazower, "Violence and the Wehrmacht in Greece, 1941–1944." I would like to thank Dr. Mazower for allowing me to read this paper and to use some of the fascinating material presented in it.

195. Engelmann, *Hitler's Germany*, p. 115.

196. Buchbender/Stertz, *Das andere Gesicht*, p. 84, letter 134.

197. *Deutsche Soldaten*, pp. 59–60.

198. *Last Letters from Stalingrad* (London, 1956), pp. 27–28.

199. Buchbender/Stertz, *Das andere Gesicht*, pp. 16–20.

200. Ibid., p. 105, letter 181.

201. Ibid., p. 112, letter 199.

202. To cite just one example from personal experience, during the Yom Kippur War of 1973 there were numerous instances of otherwise quite unobservant Israeli soldiers praying before they went into action.

203. Buchbender/Stertz, *Das andere Gesicht*, p. 146, letter 290.

204. Ibid., pp. 117–18, letter 216.

205. Bähr, *Kriegsbriefe*, pp. 327–28.

206. Ibid., p. 378.

207. E. Wolf, "Political and Moral Motives Behind the Resistance," in *The German Resistance to Hitler*, ed. H. Graml et al. (London, 1970), p. 232.

208. Buchbender/Stertz, *Das andere Gesicht*, p. 142, letter 278.

209. Ibid., pp. 142–43, letter 280.

210. Ibid., p. 143, letter 281.

211. Catholic and Lutheran chaplains in the Wehrmacht often remarked that at the front soldiers manifested much greater interest in matters religious than even only a few miles to the rear. See, e.g., BA-MA, RH27–18/174, 1.5.41–31.3.42; BA-MA, RH27–18/207, 1.10.43; BA-MA, RH27–18/178, 1.7.–30.9.42, 7.10.42, 1.1.43.

212. Buchbender/Stertz, *Das andere Gesicht*, p. 144, letter 286.

213. Ibid., p. 146, letter 290.

214. Ibid., p. 147, letter 294. Note the striking similarity between the troops' comments on the assassination attempt and Hitler's own brief radio speech on the night of 20–21 July. Cited in A. Bullock, *Hitler, A Study in Tyranny*, 2d rev. ed. (New York, 1964), pp. 749–50.

215. Engelmann, *Hitler's Germany*, p. 115.

216. Buchbender/Stertz, *Das andere Gesicht*, pp. 20–24.

217. Ibid., p. 154, letter 313.

218. Bäher, *Kriegsbriefe*, pp. 414–15.

219. Ibid., pp. 421–24.

220. Ibid., p. 403.

221. Ibid., p. 410.

222. Buchbender/Stertz, *Das andere Gesicht*, pp. 158–59, letter 323.

223. Bähr, *Kriegsbriefe*, p. 444.

224. Ibid., pp. 446–49.

225. Ibid., pp. 449–50.

226. Shils/Janowitz, *Cohesion and Disintegration*, 311–14.

227. Buchbender/Stertz, *Das andere Gesicht*, p. 160, letter 327.

228. Ibid., p. 161, letter 328.

229. Ibid., pp. 166–67, letter 342.

230. By this stage even Goebbels had to admit that the soldiers' letters were manifesting a severe deterioration of morale, and that finally Allied propaganda was beginning to have some effect. For one out of many such remarks, see *The Goebbels Diaries: The Last Days*, pp. 213–14.

231. Buchbender/Stertz, *Das andere Gesicht*, p. 167, letter 343.

232. Once more soldiers' comments were surprisingly similar to Hitler's own remarks during this final phase of his regime. See Fest, *Hitler*, pp. 724–50; also scattered throughout H. R. Trevor-Roper, *The Last Days of Hitler*, 7th rev. ed. (London, 1972); and *Hitler's Table Talk 1941–44*, intr. H. R. Trevor-Roper, 2d ed. (London, 1973).

233. Buchbender/Stertz, *Das andere Gesicht*, pp. 167–68, letter 344.

234. This is how Hillgruber, *Zweierlei Untergang*, will have it. According to Mayer, *Why Did the Heaven Not Darken?*, the war had indeed been a crusade against communism (and turned against the Jews only out of frustration at being unable to defeat Russia), but one of which he does not approve and certainly does not see as having erected a bulwark against barbarism.

235. Bullock *Hitler*, pp. 772–73.

236. See the unpublished paper by F. Stern, "Philosemitism: The White-washing of the Yellow Badge in West Germany, 1945–1952," presented at the Wiener Library Seminar, Tel-Aviv University, March 1987.

Conclusion

1. Fear of war also featured prominently in fictional accounts of the period. See, e.g., G. Orwell, *Coming Up for Air*, 2d ed. (Harmondsworth, 1962); J.-P. Sartre, *The Reprieve*, 6th ed. (New York, 1968).

2. W. Wette, "Ideologien, Propaganda und Innenpolitik als Voraussetzungen der Kriegspolitik des Dritten Reiches," in *Das Deutsche Reich und der Zweite Weltkrieg*, 1: 138–42. See also H. Auerbach, "Volksstimmung und veröffentliche Meinung in Deutschland zwischen März und November 1938," in *Machtbewusstsein in Deutschland am Vorabend des Zweiten Weltkrieges*, ed. F. Knipping and K.-J. Müller (Paderborn, 1984), pp. 274–93.

3. Engelmann, *Hitler's Germany*, p. 169.

4. Wette, *Ideologien*, p. 25.

5. Kershaw, *"Hitler Myth,"* pp. 151–68.

6. L. Niethammer, " 'Normalisierung' im Westen. Erinnerungsspuren in die 50er Jahre," in *Ist der Nationalsozialismus Geschichte?*, ed. D. Diner (Frankfurt/ M., 1987), pp. 153–84, esp. 156–58.

7. T. Mason, "The Workers' Opposition in Nazi Germany," *HWJ* 11 (1981): 120–37; S. Salter, "Class Harmony or Class Conflict?," in *Government Party and People in Nazi Germany*, ed. J. Noakes (Exeter, 1980), pp. 76–97.

8. Peukert, *Alltag und Barbarei*, esp. pp. 53–57.

9. Umbreit, *Vormachtstellung*, pp. 322–27.

10. H. Boberach (ed.), *Meldungen aus dem Reich* (München, 1968), p. 155.

11. M. G. Steinert, *Hitlers Krieg und die Deutschen* (Düsseldorf/Wien, 1971), p. 209.

12. Müller, *Das Scheitern*, pp. 939–44.

13. But one female worker in Berlin is said to have complained in October 1941 that it was quite impossible to organize resistance in the factories: "Half the workers simply haven't the courage to open their mouths. The other half seem intoxicated by all the victories and repeat whatever lies the propaganda men feed us." Engelmann, *Hitler's Germany*, p. 249.

14. A more detailed treatment of this issue in Bartov, *Missing Years*.

15. Engelmann, *Hitler's Germany*, pp. 238–39, notes that Ernst Jünger, recipient of the 1982 Goethe Prize, was engaged in executing Wehrmacht deserters in occupied Paris. Jünger is reported to have explained that though he had tried to make the executions as humane as possible, he was also "interested in observing how a person reacts to death under such circumstances," and characterized this as "a higher form of curiosity."

16. See especially U. Herbert, " 'Die guten und die schlechten Zeiten.' Überlegungen zur diachronen Analyse lebensgeschichtlicher Interviews," in *"Die Jahre weiss man nicht, wo man die heute hinsetzen soll." Faschismuserfahrungen im Ruhrgebiet*, ed. L. Niethammer (Berlin/Bonn, 1983), pp. 67–96; Niethammer, *Heimat und Front*; A. Tröger, "German Women's Memories of World War II," in *Behind the Lines: Gender and the Two World Wars*, ed. M. R. Higonnet et al. (New Haven, 1987), pp. 285–99. Also see Engelmann, *Hitler's Germany*, pp. 116– 19, for a revealing converstaion with an old friend particularly adept at suppressing unpleasant memories; and ibid., pp. 331–33, for his remark that "as soon as the first shock [of defeat] was over, the Germans allowed their recent past to sink into oblivion. At first, forgetting was a matter of self-preservation. . . . And eventually, one simply got used to acting as if certain things . . . had not happened at all."

17. See, e.g., W. Benz, "Die Abwehr der Vergangenheit. Ein Problem nur für Historiker und Moralisten?," in *Ist der Nationalsozialismus Geschichte?*, ed. D. Diner (Frankfurt/M., 1987), pp. 17–33; and ibid., H. Mommsen, "Aufarbeitung und Verdrängung. Das Dritte Reich im westdeutsche Geschichtsbewusstsein," pp. 74–88.

18. Niethammer, *Heimat und Front*, pp. 191–92.

19. Messerschmidt, *Die Wehrmacht*, pp. 334, 483, and n. 1132.

20. Mommsen, *Kriegserfahrungen*, p. 13.

21. Hillgruber, *Zweierlei Untergang*.

Works Cited

Abenheim, D. *Reforging the Iron Cross. The Search for Tradition in the West German Armed Forces.* Princeton: N.J., 1988.

Absolon, R. "Das Offizierkorps des deutschen Heeres 1935–1945." In *Das deutsche Offizierkorps 1860–1960.* Ed. H. H. Hofmann. Boppard am Rhein, 1980.

Adam, U. D. *Judenpolitik im Dritten Reich.* Düsseldorf, 1972.

Andreyev, C. *Vlasov and the Russian Liberation Movement. Soviet reality and émigré theories.* 2d ed. Cambridge, 1989.

Armstrong, J. A., ed. *Soviet Partisans in World War II.* Wisconsin, 1964.

Auerbach, H. "Volksstimmung und veröffentliche Meinung in Deutschland zwischen März und November 1938." In *Machtbewusstsein in Deutschland am Vorabend des Zweiten Weltkrieges.* Ed. F. Knipping and K.-J. Müller. Paderborn, 1984.

Bähr, W. and Bähr, H. W., eds. *Kriegsbriefe gefallener Studenten 1939–1945.* Tübingen/Stuttgart, 1952.

Baird, J. W. *The Mythical World of Nazi War Propaganda, 1939–45.* Minneapolis: Minn., 1974.

Bald, D. *Der deutsche Offizier.* München, 1984.

Baldwin, P., ed. *Reworking the Past. Hitler, the Holocaust, and the Historians' Debate.* Boston, 1990.

Balfour, M. *Propaganda in War, 1939–45.* London, 1979.

Bankwitz, P. C. F. "Maxime Weygand and the Army-Nation Concept in the Modern French Army." *FHS* 2 (1961): 157–88.

Bartov, O. "Von unten betrachtet: Überleben, Zusammenhalt und Brutalität an der Ostfront."In *Zwei Wege nach Moskau. Vom Hitler-Stalin-Pakt zum "Unternehmen Barbarossa" 1939–1941.* Ed. B. Wegner. München/Zürich, forthcoming, 1991.

219

———. "The Conduct of War: Soldiers and the Barbarization of Warfare." *JMH Special Issue: Resistance Against the National Socialist Regime* (forthcoming, 1991).

———. "The Missing Years: German Workers, German Soldiers." *GH* 8 (1990): 46–65.

———. "Daily Life and Motivation in War: The Wehrmacht in the Soviet Union." *JSS* 12 (1989): 200–14.

———. "Extremfälle der Normalität und die Normalität des Aussergewöhnlichen: Deutsche Soldaten an der Ostfront." In *Über Leben im Krieg.* Ed. U. Borsdorf and M. Jamin. Reinbeck bei Hamburg, 1989.

———. "Man and the Mass: Reality and the Heroic Image in War." *H&M* 1 (1989): 99–122.

———. "Historians on the Eastern Front: Andreas Hillgruber and Germany's Tragedy." *TAJB* 16 (1987): 325–45.

———. *The Eastern Front 1941–45, German Troops and the Barbarisation of Warfare.* London/New York, 1985/86.

Benoist-Méchin, J. *Sixty Days that Shook the West.* New York, 1963.

Benz, W. "Die Abwehr der Vergangenheit. Ein Problem nur für Historiker und Moralisten?" In *Ist der Nationalsozialismus Geschichte?* Ed. D. Diner. Frankfurt/M., 1987.

Berghahn, V. R. "Meinungsforschung im 'Dritten Reich.' " *MGM* 1 (1967): 83–119.

———. "NSDAP und 'Geistige Führung' der Wehrmacht." *VfZ* 17 (1969): 17–71.

Berning, C. *Vom "Abstammungsnachweis" zum "Zuchtwart." Vokabular des Nationalsozialismus.* Berlin, 1964.

Bertaud, J.-P. "Napoleon's Officers." *P&P* 112 (1986): 91–111.

Bien, D.D. "The Army in the French Enlightenment: Reform, Reaction and Revolution." *P&P* 85 (1979): 68–98.

"Bildserien: Verbrannte Erde; 'Partisanenbekämpfung;' eine Erschiessung sowjetischer Soldaten; sowjetische Juden als Zwangsarbeiter der Wehrmacht." In *Über Leben im Krieg.* Ed. U. Borsdorf and M. Jamin. Reinbeck bei Hamburg, 1989.

Blackburn, G. W. *Education in the Third Reich.* New York, 1959.

Bloch, M. *Strange Defeat. A Statement of Evidence Written in 1940.* New York/London, 1968.

Boberach, H., ed. *Meldungen aus dem Reich.* München, 1968.

Böll, H. *Was soll aus dem Jungen bloss werden?* Bornheim, 1981.

———. *Der Zug war pünktlich.* 12th ed. München, 1980.

Brader, D. "Foreword." To H.-U. Rudel. *Stuka Pilot.* 2d ed. Maidstone, 1973.

Bramsted, E. K. *Goebbels and National Socialist Propaganda, 1925–45.* London, 1965.

Bridenthal, R. *Mothers in the Fatherland.* New York, 1987.

———. et al. *When Biology Became Destiny.* New York, 1984.

Broszat, M. *Nationalsozialistische Polenpolitik 1939–45.* Stuttgart, 1961.

————. "Plädoyer für eine Historisierung des Nationalsozialismus." *Merkur* 435 (1985): 373–85.

————, and Fröhlich, E. *Alltag und Widerstand*. München, 1987.

Browning, C. R. *Fateful Months. Essays on the Emergence of the Final Solution*. New York, 1985.

Buchbender, O., and Stertz, R., eds. *Das andere Gesicht des Krieges*. München, 1982.

Bull, H., ed. *The Challenge of the Third Reich*. Oxford, 1986.

Bullock, A. *Hitler, A Study in Tyranny*. 2d rev. ed. New York, 1964.

Burleigh, M. *Germany Turns Eastwards. A Study of Ostforschung in the Third Reich*. 2d ed. Cambridge, 1989.

Buss, P. H. "The Non-Germans in the German Armed Forces 1939–45." Canterbury Univ. Ph.D. thesis, 1974.

Carroll, B. A. *Design for Total War*. The Hague, 1968.

Carsten, F. L. *The Reichswehr and Politics 1918–33*. 2d ed. Berkeley/Los Angeles, 1973.

————. "Germany. From Scharnhorst to Schleicher: The Prussian Officer Corps in Politics, 1806–1933." In *Soldiers and Governments*. Ed. M. Howard. London, 1957.

Cecil, R. *The Myth of the Master Race*. London, 1972.

Challener, R. D. *The French Theory of the Nation in Arms, 1866–1939*. New York, 1955.

Chapman, G. *Why France Collapsed*. London, 1968.

Chodoff, E. P. "Ideology and Primary Groups." *AFS* 9 (1983): 569–93.

Cobb, R. *French and Germans, Germans and French. A Personal Account of France under Two Occupations 1914–1918/1940–1944*. 2d ed. Hanover: N.H., 1984.

Cocks, G. "Partners and Pariahs: Jews and Medicine in Modern German Society." *LBIY* 36 (forthcoming, 1991).

————. "The Professionalization of Psychotherapy in Germany, 1928–1949." In *German Professions, 1800–1950*. Ed. G. Cocks and K. Jarausch. New York, 1990.

Cooper, M. *The Phantom War*. London, 1979.

Craig, G. A. *The Politics of the Prussian Army, 1640–1945*. 3d ed. London, 1978.

Dahms, H. G. *Die Geschichte des Zweiten Weltkrieges*. München/Berlin, 1983.

Dahrendorf, R. *Society and Democracy in Germany*. London, 1968.

Dallin, A. *German Rule in Russia 1941–45*. 2d ed. London, 1981.

Dammer, S. "Kinder, Küche, Kriegsarbeit: Die Schulung der Frauen durch die NS-Frauenschaft." In *Mutterkreuz und Arbeitsbuch*. Ed. Frauengruppe Faschismusforschung. Frankfurt/M., 1981.

Davis, G. H. "Prisoners of War in Twentieth-Century War Economies." *JCH* 12 (1977): 623–34.

Dawidowicz, L. *The War against the Jews 1933–45*. 3d ed. Harmondsworth, 1979.

"Debate: Germany, 'Domestic Crisis' and War in 1939." Comments by D. Kaiser and T. W. Mason, reply by R. J. Overy. *P&P* 122 (1989): 200–40.

Deist, W. *The Wehrmacht and German Rearmament*. London/Basingstoke, 1981.

———. "Die Aufrüstung der Wehrmacht," In *Ursachen und Voraussetzungen der deutschen Kriegspolitik*. Stuttgart, 1979. Vol. 1 of *Das Deutsche Reich und der Zweite Weltkrieg*.

Demeter, K. *The German Officer Corps in Society and State, 1650–1945*. London, 1965.

Deutsch, H. C. *Hitler and his Generals*. Minneapolis: Minn., 1974.

———. *The Conspiracy against Hitler in the Twilight War*. Minneapolis: Minn., 1968.

Deutsche Soldaten sehen die Sowjetunion. Berlin, 1941.

Diner, D. "Rassistisches Völkerrecht: Elemente einer nationalsozialistischen Weltordnung." VfZ 1 (1989): 23–56.

Doenitz, K. *Memoirs*. 2d ed. London, 1959.

Eichholz, D. "Der 'Generalplan Ost.' " *JfG* 26 (1982): 217–74.

Ellis, J. *The Social History of the Machine Gun*. 2d ed. London, 1987.

Ellis, L. F. *The War in France and Flanders, 1939–40*. London, 1953.

Engelmann, B. *In Hitler's Germany. Daily Life in the Third Reich*. New York, 1986.

Erickson, J. *The Road to Stalingrad*. 2d ed. London, 1985. Vol. 1 of *Stalin's War with Germany*.

Evans, R. J. *In Hitler's Shadow*. London, 1989.

Fest, J. *Hitler*. 3d ed. Harmondsworth, 1982.

Flessau, K.-I., et al., eds. *Erziehung im Nationalsozialismus*. Köln, 1987.

Flottmann, J., and Möller-Witten, H. *Opfergang der Generale. Die Verluste der Generale und Admirale und der im gleichen Dienstrang stehenden sonstigen Offiziere und Beamten im Zweiten Weltkrieg*. Berlin, 1952.

Focke, H., and Strocka, M. *Alltag der Gleichgeschalteten*. Reinbek bei Hamburg, 1985.

Förster, J. "The German Army and the Ideological War against the Soviet Union." In *The Policies of Genocide*. Ed. G. Hirschfeld. London, 1986.

———. "New Wine in Old Skins? The Wehrmacht and the War of 'Weltanschauungen,' 1941." In *The German Military in the Age of Total War*. Ed. W. Deist. Leamington Spa/New Hampshire, 1985.

———. "Hitlers Entscheidung für den Krieg gegen die Sowjetunion;" "Das Unternehmen 'Barbarossa' als Eroberungs- und Vernichtungskrieg;" (with G. R. Ueberschär), "Freiwillige für den 'Kreuzzug Europas gegen den Bolschewismus.'" All in *Der Angriff auf die Sowjetunion*. Stuttgart, 1983. Vol. 4 of *Das Deutsche Reich und der Zweite Weltkrieg*.

———. " 'Croisade de l'Europe contre le Bolchevisme.' " *RHDGM* 30 (1980): 1–26.

Foy, D. A. *For You the War Is Over. American Prisoners of War in Nazi Germany*. New York. 1984.

Frei, N. *Der Führerstaat. Nationalsozialistische Herrschaft 1933 bis 1945*. München, 1987.

Friedländer, S. "Some Reflections on the Historisation of National Socialism." *TAJB* 16 (1987): 310–24.

Fuller, J. F. C. *The Decisive Battles of the Western World.* 2d ed. London, 1970. Vol. 1.

Ganssmüller, C. *Die Erbgesundheitspolitik des Dritten Reiches.* Köln/Wien, 1987.

Gelber, Y. "Military Education and Ideology." *Ma'arakhot* 267 (1979): 8–12 (Hebrew).

Geuter, U. *Die Professionalisierung der deutschen Psychologie im Nationalsozialismus.* 2d ed. Frankfurt/M., 1988.

Geyer, M. *Aufrüstung oder Sicherheit. Die Reichswehr in der Krise der Machtpolitik 1924–36.* Wiesbaden, 1980.

———. "Etudes in Political History: Reichswehr, NSDAP, and the Seizure of Power." In *The Nazi Machtergreifung.* Ed. P. D. Stachura. London, 1983.

———. "The Militarization of Europe, 1914–1945." In *The Militarization of the Western World.* Ed. J. R. Gillis. New Brunswick/London, 1989.

———. "Professionals and Junkers: German Rearmament and Politics in the Weimar Republic." In *Social Change and Political Development in Weimar Germany.* Ed. R. Bessel and E. J. Feuchtwanger. London, 1981.

Giles, G. J. *Students and National Socialism in Germany.* Princeton: N.J., 1985.

The Goebbels Diaries 1939–1941. Trans. and ed. F. Taylor, 2d ed. Harmondsworth, 1984.

The Goebbels Diaries: The Last Days. Ed. H. Trevor-Roper, 2d ed. London/Sydney, 1979.

Gordon, S. *Hitler, Germans, and the "Jewish Question."* Princeton: N.J., 1984.

Goutard, A. *The Battle of France, 1940.* London, 1958.

Graml, H., et al. *The German Resistance to Hitler.* London, 1970.

Gruchmann, L. " 'Blutschutzgesetz' und Justiz. Zur Entstehung und Auswirkung des Nürnberger Gesetzes vom 15. September 1935." *VfZ* 31 (1983): 418–42.

Guderian, H. *Panzer Leader.* 3d ed. London, 1977.

Gurfein M. I., and Janowitz, M. "Trends in Wehrmacht Morale." In *Propaganda in War and Crisis.* Ed. D. Lerner. New York, 1951.

Güstow, D. *Tödlicher Alltag. Strafverteidiger im Dritten Reich.* Berlin, 1981.

Habe, H. *A Thousand shall Fall.* London, 1942.

Habermas, J. "Eine Art Schadensabwicklung." *Die Zeit,* 11.7.86.

Halder, F. *Kriegstagebuch 1939–42.* Ed. H.-A. Jacobsen. 3 vols. Stuttgart, 1962–64.

Hastings, M. *Das Reich. Resistance and the March of the 2nd SS Panzer Division Through France, June 1944.* London, 1981.

Hayes, P. *Industry and Ideology. IG Farben in the Nazi Era.* 2d ed. Cambridge, 1989.

Heck, A. *A Child of Hitler.* 3d ed. Toronto/New York/London, 1986.

Heiber, H. "Der Generalplan Ost." *VfZ* (1958): 281–325.

Herbert, U. *Fremdarbeiter: Politik und Praxis des "Ausländer-Einsatzes" in der Kriegswirtschaft des Dritten Reiches.* Berlin/Bonn, 1985.

———. " 'Die guten und die schlechten Zeiten.' Überlegungen zur diachronen Analyse lebensgeschichtlicher Interviews." In *Die Jahre weiss man nicht, wo man die heute hinsetzen soll." Faschismuserfahrungen im Ruhrgebiet*. Ed. L. Niethammer. Berlin/Bonn, 1983.

Herf, J. *Reactionary Modernism*. 2d ed. Cambridge, 1986.

Herwarth, J. von. *Against Two Evils*. London, 1981.

Hilberg, R. *The Destruction of European Jews*. Rev. ed. New York, 1986.

Hillgruber, A. *Zweierlei Untergang. Die Zerschlagung des Deutschen Reiches und das Ende des europäischen Judentums*. Berlin, 1986.

———. *Hitlers Strategie. Politik und Kriegführung 1940–41*. Frankfurt/M., 1965.

Hirsch M., et al. eds. *Recht, Verwaltung und Justiz im Nationalsozialismus*. Köln, 1984.

Hirschfeld, G., ed. *The Policies of Genocide. Jews and Soviet Prisoners of War in Nazi Germany*. London/Boston/Sydney, 1986.

"Historikerstreit." Die Dokumentation der Kontroverse um die Einzigartigkeit der nationalsozialistischen Judenvernichtung. Ed. Serie Piper. 3d ed. München, 1987.

Hitler's Table Talk 1941–44. Intr. H. R. Trevor-Roper. 2d ed. London, 1973.

Hoffmann, J. "Die Sowjetunion bis zum Vorabend des deutschen Angriffs;" "Die Kriegführung aus der Sicht der Sowjetunion." Both in *Der Angriff auf die Sowjetunion*. Stuttgart, 1983. Vol. 4 of *Das Deutsche Reich und der Zweite Weltkrieg*.

———. *Deutsche und Kalmyken 1942 bis 1945*. Freiburg, 1977.

———. *Die Ostlegionen 1941–43*. Freiburg, 1977.

Hoffmann, P. *German Resistance to Hitler*. Cambridge: Mass., 1988.

Holmes, R. *Firing Line*. 2d ed. Harmondsworth, 1987.

Homze, E. L. *Foreign Labour in Nazi Germany*. Princeton: N.J., 1967.

Horne, A. *To Lose a Battle*. Harmondsworth, 1979.

Howard, M. *The Franco-Prussian War*. 3d ed. London, 1981.

Huebsch, N. A. "The 'Wolf Cubs' of the New Order: The Indoctrination and Training of the Hitler Youth." In *Nazism and the Common Man*. Ed. O. C. Mitchell. Washington D. C., 1981.

Hull, D. S. *Film in the Third Reich*. Berkeley: California, 1969.

Hüppauf, B. "Langemarck, Verdun and the Myth of a New Man in Germany after the First World War." *W&S* 6 (1988): 70–103.

Irving, D. *The Rise and Fall of the Luftwaffe*. London, 1976.

Jäckel, E. *Hitlers Herrschaft: Vollzug einer Weltanschauung*. Stuttgart, 1986.

———. *Hitlers Weltanschauung: Entwurf einer Herrschaft*. 3d ed. Stuttgart, 1986.

———. *Frankreich in Hitlers Europa*. Stuttgart, 1966.

Jacobsen, H.-A. "Kommissarbefehl und Massenexekutionen sowjetische Kriegsgefangener." In *Anatomie des SS-Staates*. Ed. H. Buchheim et al. Olten, 1965.

Jünger, E. *Strahlungen II*. In *Tagebücher III*. Vol. 3 of *Sämtliche Werke*. Stuttgart, 1979 [1949].

Kaes, A. *From Hitler to Heimat: The Return of History as Film*. Cambridge: Mass., 1989.

Kaiser, D. *Economic Diplomacy and the Origins of the Second World War*. Princeton: N.J., 1980.

Kater, M. *Doctors under Hitler*. Chapel Hill: N.C., 1989.

Keegan, J. *The Face of Battle*. 2d ed. London, 1976.

Kellet, A. *Combat Motivation*. Boston, 1982.

Kershaw, I. *The Nazi Dictatorship*. 2d ed. London, 1989.

———. "Neue deutsche Unruhe? Das Ausland und der Streit um die deutsche National- und Zeitgeschichte." In *Streitfall Deutsche Geschichte*. Ed. Landeszentrale für politische Bildung Nordrhein-Westfalen. 1989.

———. *The "Hitler Myth." Image and Reality in the Third Reich*. Oxford, 1987.

———. *Popular Opinion and Political Dissent in the Third Reich*. Oxford, 1983.

———. "How Effective was Nazi Propaganda?" In *Nazi Propaganda*. Ed. D. Welch. London, 1983.

———. "Ideology, Propaganda, and the Rise of the Nazi Party." In *The Nazi Machtergreifung*. Ed. P. D. Stachura. London, 1983.

Kesselring. *The Memoirs*. London, 1953.

Kitchen, M. *The German Officer Corps, 1890–1914*. London, 1968.

———. *The Silent Dictatorship. The Politics of the German High Command under Hindenburg and Ludendorff, 1916–1918*. New York, 1976.

Klee, E. *Was sie taten—was sie wurden. Ärtze, Juristen und andere Beteiligte am Kranken- oder Judenmord*. Frankfurt/M., 1986.

———. *Dokumente zur "Euthanasie."* Frankfurt/M., 1985.

———. *"Euthanasie" im NS Staat. Die "Vernichtung lebensunwerten Lebens."* Frankfurt/M., 1983.

———. et al., eds. *"Schöne Zeiten." Judenmord aus der Sicht der Täter und Gaffer*. Frankfurt/M., 1988.

Klemperer, V. *Die Unbewältigte Sprache*. 3d ed. Darmstadt, n.d.

Klink, E. *Das Gesetz des Handelns*. Stuttgart, 1966.

———. "Die militärische Konzeption des Krieges gegen die Sowjetunion: Die Landkriegführung"; "Die Operationsführung: Heer und Kriegsmarine." Both in *Der Angriff auf die Sowjetunion*. Stuttgart, 1983. Vol. 4 of *Das Deutsche Reich und der Zweite Weltkrieg*.

Klönne, A. *Jugend im Dritten Reich*. 2d ed. Köln, 1984.

Koch, H. W. *The Hitler Youth*. London, 1975.

König, S. *Vom Dienst am Recht: Rechtsanwälte als Strafverteidiger im Nationalsozialismus*. New York, 1987.

Krausnick, H. "Kommissarbefehl und 'Gerichtsbarkeitserlass Barbarossa' in neuer Sicht." *VfZ* 25 (1977): 682–758.

———, and Wilhelm, H.-H. *Die Truppe des Weltanschauungskrieges*. Stuttgart, 1981.

———, and Deutsch, H. C., eds. *Helmut Groscurth: Tagebücher eines Abwehroffiziers*. Stuttgart, 1970.

Kriegstagebuch des Oberkommando der Wehrmacht 1940–45. Ed. P. E. Schramm. 4 vols. Frankfurt/M., 1961–65.

Kroener, B. R. "Auf dem Weg zu einer 'nationalsozialistischen Volksarmee.' Die soziale Öffnung des Heeresoffizierkorps im Zweiten Weltkrieg." In *Von Stalingrad zur Währungsreform. Zur Sozialgeschichte des Umbruchs in Deutschland*. Ed. M. Broszat et al. München, 1988.

———. "Die Personellen Ressourcen des Dritten Reiches im Spannungsfeld zwischen Wehrmacht, Bürokratie und Kriegswirtschaft 1939–1942." In *Organisation und Mobilisierung des deutschen Machtbereichs*. Stuttgart, 1988. Vol. 5/1 of *Das deutsche Reich und der Zweite Weltkrieg*.

———. "Squaring the Circle. Blitzkrieg Strategy and the Manpower Shortage, 1939–42." In *The German Military in the Age of Total War*. Ed. W. Deist. Leamington Spa/New Hampshire, 1985.

Kühnrich, H. *Der Partisanenkrieg in Europa, 1939–45*. Berlin, 1965.

Kulka, O. D. "Die Nürnberger Rassengesetze und die deutsche Bevölkerung im Lichte geheimer NS-Lage- und Stimmungsberichte." *VfZ* 32 (1984): 582–624.

Kupffer, H. *Der Faschismus und das Menschenbild der deutschen Pädagogik*. Frankfurt/M., 1984.

Kurze, H. "Das Bild des Offiziers in der deutschen Literatur." In *Das deutsche Offizierkorps, 1860–1960*. Ed. H. H. Hofmann. Boppard am Rhein, 1980.

Last Letters from Stalingrad. London, 1956.

Le Roy Ladurie, E. *Carnival in Romans*. 3d ed. Harmondsworth, 1981.

Leed, E. J. *No Man's Land*. Cambridge, 1979.

Liddell Hart, B. H. *History of the First World War*. 3d ed. London, 1979.

———. *The Other Side of the Hill*. London, 1948.

Loewenberg, P. "The Psychological Origins of the Nazi Youth Combat." *AHR* 76 (1971): 1457–1502.

Magenheimer, H. *Abwehrschlachten an der Weichsel 1945*. Freiburg, 1976.

Maier, C. S. *The Unmasterable Past*. Cambridge: Mass., 1988.

Majer, D. *"Fremdvölkische" im Dritten Reich*. Boppard am Rhein, 1981.

Mann, E. *School for Barbarians*. London, 1939.

Manstein, E. von. *Verlorene Siege*. 2d ed. Frankfurt/M., 1964.

———. *Aus einem Soldatenleben*. Bonn, 1958.

Marshall, S. L. A. *Men Against Fire*. New York, 1947.

Mason, T. W., ed. "The Workers' Opposition in Nazi Germany." *HWJ* 11 (1981): 120–37.

———. *Sozialpolitik im Dritten Reich*. Opladen, 1977.

———. "Women in Germany, 1925–40: Family, Welfare and Work." Parts 1–2, *HWJ* 1 (1976): 74–113; 2 (1976): 5–32.

———. *Arbeiterklasse und Volksgemeinschaft*. Opladen, 1975.

———. "Innere Krise und Angriffskrieg." In *Wirtschaft und Rüstung am Vorabend des Zweiten Weltkrieges*. Ed. F. Forstmeier and H.-E. Volkmann. Düsseldorf, 1975.

Mayer, A. J. *Why Did the Heaven Not Darken?* New York, 1988.

Mazower, M. "Violence and the Wehrmacht in Greece, 1941–1944." Unpublished paper.

McIntyre, S. J. "Women and the Professions in Germany 1930–40." In *German*

Democracy and the Triumph of Hitler. Ed. A. Nicholls and E. Matthias. London, 1971.

Meier-Welcker, H. *Aufzeichnungen eines Generalstabsoffiziers 1939–42.* Freiburg, 1982.

Merson, A. *Communist Resistance in Nazi Germany.* London, 1987.

Messerschmidt, M. "German Military Law in the Second World War." In *The German Military in the Age of Total War.* Ed. W. Deist. Leamington Spa/ New Hampshire, 1985.

———. "Deutsche Militärgerichtsbarkeit im Zweiten Weltkrieg." In *Recht, Verwaltung und Justiz im Nationalsozialismus.* Ed. M. Hirsch. Köln, 1984.

———. "The Wehrmacht and the Volksgemeinschaft." *JCH* 18 (1983): 719–40.

———. "Kommandobefehl und NS-Völkerrechtsdenken." *RDP* 11 (1972): 110–34.

———. *Die Wehrmacht im NS-Staat. Zeit der Indoktrination.* Hamburg, 1969.

———. "Werden und Prägung des preussischen Offizierkorps—Ein Überblick." In *Offiziere im Bild vom Dokumente aus drei Jahrhunderten.* Stuttgart, 1964.

———, and Wüllner, F. *Die Wehrmachtjustiz im Dienste des Nationalsonzialismus. Zerstörung einer Legende.* Baden-Baden, 1987.

Milward, A. S. *The German Economy at War.* London, 1965.

Mommsen, H. "Kriegserfahrungen." In *Über Leben im Krieg.* Ed. U. Borsdorf and M. Jamin. Reinbeck bei Hamburg, 1989.

———. "Aufarbeitung und Verdrängung. Das Dritte Reich im westdeutsche Geschichtsbewusstsein." In *Ist der Nationalsozialismus Geschichte?* Ed. D. Diner. Frankfurt/M., 1987.

Moran, Lord. *The Anatomy of Courage.* 2d ed. London, 1966.

Mosse, G. L. *Fallen Soldiers. Reshaping the Memory of the World Wars.* New York, 1990.

Mueller-Hillebrand, B. *Das Heer.* 3 vols. Darmstadt, 1954; Frankfurt/M., 1956/ 1969.

Müller, K.-J. *The Army, Politics and Society in Germany, 1933–45.* New York, 1987.

———. *General Ludwig Beck.* Boppard am Rhein, 1980.

———. *Das Heer und Hitler.* Stuttgart, 1969.

Müller, N. "Dokumente zur Rolle der Wehrmacht bei der Deportation sowjetischer Bürger zur Zwangsarbeit in Deutschland 1941–44." *BAZW* 4 (1970): 29–62.

Müller, R.-D. *Das Tor zur Weltmacht.* Boppard am Rhein, 1984.

———. "Von der Wirtschaftsallianz zum kolonialen Ausbeutungskrieg;" "Das Scheitern der wirtschaftlichen 'Blitzkriegstrategie." Both in *Der Angriff auf die Sowjetunion.* Stuttgart, 1983. Vol. 4 of *Das Deutsche Reich und der Zweite Weltkrieg.*

———. "Industrielle Interessenpolitik im Rahmen des 'Generalplan Ost.'" *MGM* 1 (1981): 101–41.

Mulligan, T. P. "Reckoning the Cost of People's War: the German Experience in the Central USSR." *RH* 5 (1982): 27–48.

Mushaben, J. M. "Youth Protest and the Democratic State: Reflections on the Rise of Anti-Political Culture in Prewar Germany and the German Federal Republic." *RPS* 2 (1986): 171–97.

Niethammer, L. " 'Normalisierung' im Westen. Erinnerungsspuren in die 50er Jahre." In *Ist der Nationalsozialismus Geschichte?* Ed. D. Diner. Frankfurt/ M., 1987.

———. "Heimat und Front. Versuch, zehn Kriegserinnerungen aus der Arbei- terklasse des Ruhrgebietes zu verstehen." In *"Die Jahre weiss man nicht, wo man die heute hinsetzen soll." Faschismus im Ruhrgebiet.* Ed. L. Niethammer. Berlin, 1983.

———. "Anmerkungen zur Alltagsgeschichte." In *Geschichte im Alltag—Alltag in der Geschichte.* Ed. K. Bergmann and R. Schörken. Düsseldorf, 1982.

Nolte, E. *Das Vergehen der Vergangenheit.* Berlin/Frankfurt/M., 1987.

———. "Vergangenheit, die nicht vergehen will." *FAZ*, 6.6.86.

———. "Between Myth and Revisionism?" In *Aspects of the Third Reich.* Ed. W. Koch. London, 1985.

Nyssen, E. *Schule im Nationalsozialismus.* Heidelberg, 1979.

O'Neill, R. J. *The German Army and the Nazi Party, 1933–39.* London, 1966.

Orwell, G. *Coming Up for Air.* 2d ed. Harmondsworth, 1962.

Overy, R. J. "Germany, 'Domestic Crisis' and War in 1939." *P&P* 116 (1987): 138–68.

———. *Goering: The "Iron Man".* London, 1984.

———. "Hitler's War and the German Economy: A Reinterpretation." *EHR* 35 (1982): 272–91.

Papke, G. "Offizierkorps und Anciennität." In *Untersuchungen zur Geschichte des Offizierkorps.* Ed. H. Meier-Welcker. Stuttgart, 1962.

Paul, W. *Geschichte der 18. Panzer Division 1940–43.* Freiburg, n.d.

Pauwels, J. R. *Women, Nazis, and Universities.* Westport: Conn., 1984.

Paxton, R. O. *Vichy France.* 2d ed. New York, 1975.

Petzina, D. "Die Mobilisierung Deutscher Arbeitskräfte vor und während des Zweiten Weltkrieges." *VfZ* 4 (1970): 443–55.

Peukert, D. J. K. "Alltag und Barbarei." In *Ist der Nationalsozialismus Geschichte?* Ed. D. Diner. Frankfurt/M., 1987.

———. *Inside Nazi Germany.* London, 1987.

———, and Bajohr, F. *Spuren des Widerstands: Die Bergarbeiterbewegung im Dritten Reich und im Exil.* München, 1987.

Pfahlmann, H. *Fremdarbeiter und Kriegsgefangene in der deutschen Kriegswirt- schaft 1939–45.* Darmstadt, 1968.

Planter, G., ed. *Schule im Dritten Reich—Erziehung zum Tod?* 2d ed. München, 1984.

Ploetz. *Geschichte des Zweiten Weltkrieges.* 2d ed. Würzburg, 1960.

Quinnett, R. L. "The National Socialist Leadership Officers." Univ. of Oklahoma Ph.D. thesis, 1973.

Redelis, V. *Partisanenkrieg.* Heidelberg, 1958.

Reinhardt, K. *Die Wende vor Moskau.* Stuttgart, 1972.

Richardson, F. M. *Fighting Spirit*. London, 1978.

Richardson, H. F., ed. *Sieg Heil! War Letters of Tank Gunner Karl Fuchs 1937–1941*. Hamden: Conn., 1987.

Ringer, F. "Bildung, Wirtschaft und Gesellschaft in Deutschland 1800–1960." *GuG* 6 (1980): 5–35.

Ritter, G. *Carl Goerdeler und die deutsche Widerstands-Bewegung*. Stuttgart, 1954.

Rohde, H. "Hitlers erster 'Blitzkrieg' und seine Auswirkungen auf Nordosteuropa." In *Die Errichtung der Hegemonie auf dem Europäischen Kontinent*. Stuttgart, 1979. Vol. 2 of *Das Deutsche Reich und der Zweite Weltkrieg*.

———. *Das Deutsche Wehrmachttransportwesen im Zweiten Weltkrieg*. Stuttgart, 1971.

Rosas, A. *The Legal Status of Prisoners of War*. Helsinki, 1976.

Ruck, M. *Bollwerk gegen Hitler? Arbeiterschaft, Arbeiterbewegung und die Anfänge des Nationalsozialismus*. Köln, 1988.

Rudel, H.-U. *Stuka Pilot*. 2d ed. Maidstone, 1973.

Rüthers, B. *Entartetes Recht: Rechtslehren und Kronjuristen im Dritten Reich*. München, 1988.

Sajer, G. *The Forgotten Soldier*. 2d ed. London, 1977.

Salter, S. "Class Harmony or Class Conflict?" In *Government Party and People in Nazi Germany*. Ed. J. Noakes. Exeter, 1980.

Sartre, J.-P. *Les carnets de la drôle de guerre*. Paris, 1983.

———. *The Reprieve*. 6th ed. New York, 1968.

Schleunes, K. A. *The Twisted Road to Auschwitz. Nazi Policy toward German Jews, 1933–1939*. Urbana/Chicago/London, 1970.

Schmuhl, H. W. *Rassenhygiene, Nationalsozialismus, Euthanasie*. Göttingen, 1987.

Schoenbaum, D. *Hitler's Social Revolution*. New York/London, 1966.

Scholtz, H. *Erziehung und Unterricht unterm Hakenkreuz*. Göttingen, 1985.

Schörken, R. "Jugendalltag im Dritten Reich." In *Geschichte im Alltag—Alltag in der Geschichte*. Ed. K. Bergmann and R. Schörken. Düsseldorf, 1982.

Schulte, T. *The German Army and Nazi Policies in Occupied Russia*. Oxford/New York/Munich, 1989.

Schustereit, H. *Vabanque: Hitlers Angriff auf die Sowjetunion 1941 als Versuch, durch den Sieg im Osten den Westen zu bezwingen*. Herford: F.R.G., 1988.

Schweling, O. P. *Die deutsche Militärjustiz in der Zeit des Nationalsozialismus*. 2d ed. Marburg, 1978.

Seaton, A. *The Russo-German War 1941–45*. London, 1971.

Seidler, F. *Prostitution, Homosexualität, Selbstverstümmelung. Probleme der deutsche Sanitätsführung 1939–45*. Neckargemünd, 1977.

Shils, E. A. and Janowitz, M. "Cohesion and Disintegration in the Wehrmacht in World War II." *POQ* 12 (1948): 280–315.

Spaeter, H. and Schramm, W. Ritter von. *Die Geschichte des Panzerkorps Grossdeutschland*. 3 vols. Bielefeld, 1958.

Speer, A. *Inside the Third Reich*. 5th ed. London, 1979.

Stachura, P. D. *Nazi Youth in the Weimar Republic*. Santa Barbara: Calif., 1975.

Steinert, M. G. *Hitler's War and the Germans*. Athens: Ohio, 1977.

Stephens, F. J. *Hitler Youth*. London, 1973.

Stephenson, J. " 'Emancipation' and its Problems: War and Society in Württemberg 1939–45." *EHQ* 17 (1987): 345–65.

———. *Women in Nazi Society*. London, 1975.

Stern, F. "Philosemitism: The Whitewashing of the Yellow Badge in West Germany, 1945–1952." Unpublished paper.

Stern, F. *Dreams and Delusions: The Drama of German History*. London, 1988.

Stevenson, J. *British Society 1914–45*. Harmondsworth, 1984.

Streim, A. *Die Behandlung sowjetischer Kriegsgefangener in "Fall Barbarossa"*. Heidelberg, 1981.

Streit, C. "The German Army and the Policies of Genocide." In *The Policies of Genocide*. Ed. G. Hirschfeld. London, 1986.

———. *Keine Kameraden*. Stuttgart, 1978.

Stumpf, R. *Die Wehrmacht-Elite. Rang- und Herkunftsstruktur der deutschen Generale und Admirale 1933–45*. Boppard am Rhein, 1982.

Stürmer, M. "Geschichte in geschichtslosem Land." *FAZ*, 25.4.86.

Sydnor, C. W., Jr. *Soldiers of Destruction. The SS Death's Head Division, 1933–1945*. Princeton: N.J., 1977.

Sywottek, J. *Mobilmachung für den Krieg*. Opladen, 1976.

Tal, U. *"Political Faith" of Nazism Prior to the Holocaust*. Tel-Aviv, 1978.

Tamir, A. "Quality Versus Quantity." *Ma'arakhot* 250 (1976): 8–12, 38 (Hebrew).

Tenfelde, K. "Schwierigkeiten mit dem Alltag." *GuG* 10 (1984): 376–94.

Tessin, G. *Formationsgeschichte der Wehrmacht, 1933–39*. Boppard am Rhein, 1959.

Thompson, P. *The Voice of the Past: Oral History*. Oxford, 1978.

Topitsch, E. *Stalin's War*. New York, 1987.

Tournier, M. *The Ogre*. New York, 1972.

Trevor-Roper, H. R. *The Last Days of Hitler*. 7th rev. ed. London, 1972.

Tröger, A. "German Women's Memories of World War II." In *Behind the Lines: Gender and the Two World Wars*. Ed. M. R. Higonnet et al. New Haven, 1987.

Turner, H. A., Jr. "Fascism and Modernization." In *Reappraisals of Fascism*. Ed. H. A. Turner, Jr. New York, 1975.

Ueberschär, G. R. and Wette, W., eds. *"Unternehmen Barbarossa"*. Paderborn, 1984.

Umbreit, H. "Der Kampf um die Vormachtstellung in Westeuropa." In *Die Errichtung der Hegemonie auf dem Europäischen Kontinent*. Stuttgart, 1979. Vol 2 of *Das Deutsche Reich und der Zweite Weltkrieg*.

———. *Deutsche Militärverwaltung 1938/39*. Stuttgart, 1977.

———. *Der Militärbefehlshaber in Frankreich 1940–44*. Boppard am Rhein, 1968.

van Creveld, M. *Fighting Power*. Westport: Conn., 1982.

———. *Supplying War. Logistics from Wallenstein to Patton*. 3d ed. New York, 1980.

Volkmann, H.-E. "Die NS-Wirtschaft in Vorbereitung des Krieges." in *Ursachen und Voraussetzungen der deutschen Kriegspolitik.* Stuttgart, 1979. Vol. 1 of *Das Deutsche Reich und der Zweite Weltkrieg.*

Walzer, M. *Just and Unjust Wars.* 3d ed. Harmondsworth, 1980.

Weber, R. G. S. *The German Student Corps in the Third Reich.* London, 1986.

Wegner, B. "Auf dem Wege zur Pangermanischen Armee." *MGM* 2 (1980): 101–36.

———. "Das Führerkorps der Waffen-SS im Kriege." In *Das deutsche Offizierkorps, 1860–1960.* Ed. H. H. Hofmann. Boppard am Rhein, 1980.

Wehler, H.-U. *Entsorgung der deutschen Vergangenheit?* München, 1988.

Weinberg, G. L. "Dokumentation: Adolf Hitler und der NS-Führungsoffizier (NSFO)." *VfZ* 12 (1964): 443–56.

Weingartner, J. J. *Crossroads of Death. The Story of the Malmédy Massacre and Trial.* Berkeley/Los Angeles, 1979.

Welch, D., ed. *Nazi Propaganda.* London, 1983.

———. "Nazi Wartime Newsreel Propaganda." In *Film & Radio Propaganda in World War II.* Ed. K. R. M. Short. Knoxville: Tenn., 1983.

Westphal, S. *The German Army in the West.* London, 1951.

Wette, W. "Ideologien, Propaganda und Innenpolitik als Voraussetzungen der Kriegspolitik des Dritten Reiches." In *Ursachen und Voraussetzungen der deutschen Kriegspolitik.* Stuttgart, 1979. Vol. 1 of *Das Deutsche Reich und der Zweite Weltkrieg.*

Wheeler-Bennett, J. *The Nemesis of Power.* 2d ed. London, 1980.

Winkler, D. *Frauenarbeit im "Dritten Reich."* Hamburg, 1977.

Winter, J. M. *The Great War and the British People.* Cambridge: Mass., 1986.

Wippermann, W. "Fascism and the History of Everyday Life." Unpublished paper.

Wolf, E. "Political and Moral Motives Behind the Resistance." In *The German Resistance to Hitler.* Ed. H. Graml et al. London, 1970.

Woloch, I. "Napoleonic Conscription: State Power and civil Society." Unpublished paper.

Zeman, Z. A. B. *Nazi Propaganda.* London, 1964.

Zieger, W. *Das deutsche Heeresveterinärwesen im Zweiten Weltkrieg.* Freiburg, 1973.

Zitelmann, R. *Hitler: Selbstverständnis eines Revolutionärs.* Stuttgart, 1987.

Index